T0230689

Resource Recovery from Municipal Sewage Plants

An Energy-Water-Nutrients Nexus for Developing Countries

Resource Recovery from Municipal Sewage Plants

An Energy-Water-Nutrients Nexus for Developing Countries

Musaida Mercy Manyuchi, Charles Mbohwa, and Edison Muzenda

CRC Press
Taylor & Francis Group
Boca Raton London New York

CRC Press is an imprint of the
Taylor & Francis Group, an **informa** business

CRC Press
Taylor & Francis Group
6000 Broken Sound Parkway NW, Suite 300
Boca Raton, FL 33487-2742

© 2019 by Taylor & Francis Group, LLC
CRC Press is an imprint of Taylor & Francis Group, an Informa business

No claim to original U.S. Government works

Printed on acid-free paper

International Standard Book Number-13: 978-1-138-58400-6 (Hardback)
International Standard Book Number-13: 978-0-4295-0623-9 (eBook)

Visit the Taylor & Francis Web site at
http://www.taylorandfrancis.com

and the CRC Press Web site at
http://www.crcpress.com

Contents

List of Figures, xiii

List of Tables, xv

Preface, xvii

Authors, xix

Abbreviations, xxiii

CHAPTER 1 ▪ Resource Recovery from Municipal
Plants: A General Review   1

  1.1 BACKGROUND   1

  1.2 MOTIVATION   2

CHAPTER 2 ▪ Municipal Sewage Wastewater Treatment   5

  2.1 SEWAGE WASTEWATER TREATMENT   5

  2.2 SEWAGE WASTEWATER CHARACTERISTICS   6

      2.2.1 Fecal Coliform Content   7

      2.2.2 E. coli Content   7

      2.2.3 Temperature   8

      2.2.4 Organic Material in Raw Sewage Wastewater 8

          2.2.4.1 Municipal Sewage BOD_5 8

          2.2.4.2 Municipal Sewage COD   8

      2.2.5 Municipal Sewage Total Nitrogen Content   9

      2.2.6 Sewage Total Phosphates Content   9

2.2.7   Sewage pH   9

2.2.8   Sewage Color and Odor   9

2.2.9   Sulfate Concentration   10

2.2.10 Sewage Solids   10

2.2.11 Chlorides   10

2.2.12 Toxic Metals and Compounds   10

2.3   NUTRIENT REMOVAL IN MUNICIPAL
      SEWAGE USING BIOAUGUMENTATION   10

2.3.1   Primary Sewage Treatment   11

2.3.2   Secondary Sewage Treatment   11

2.3.3   Tertiary Sewage Treatment   12

2.4   MUNICIPAL SEWAGE TREATMENT
      CONDITIONS   12

2.4.1   Aerobic Conditions   12

2.4.2   Anaerobic Conditions   13

2.4.3   Anoxic Conditions   13

2.4.4   Anammox Conditions   13

CHAPTER 3 ▪ Anaerobic Treatment of Municipal
              Wastewater Bioaugmentation   15

3.1   INTRODUCTION   15

3.2   HYCURA BIOCHEMICAL PROPERTIES   16

3.3   USE OF HYCURA IN MUNICIPAL
      WASTEWATER TREATMENT   17

3.4   HYCURA KINETICS DURING MUNICIPAL
      WASTEWATER TREATMENT   19

CHAPTER 4 ▪ Resource Recovery from Municipal
              Sewage   21

4.1   TREATED SEWAGE EFFLUENT   21

4.2   BIONUTRIENT RECOVERY   22
4.3   BIOGAS FROM SEWAGE SLUDGE   23

    4.3.1   Conditions for Biogas Production   23

    4.3.2   Biogas Generation from Municipal
            Wastewater Using Hycura Bioaugmentation   26

    4.3.3   Benefits of Harnessing Biogas from
            Municipal Plants   26

        4.3.3.1   *Alternative and Renewable Energy
                Source*   26

        4.3.3.2   *Reduced Greenhouse Emissions*   26

        4.3.3.3   *Reduced Dependence on Fossil Fuels* 28

        4.3.3.4   *Waste Reduction and Management*   28

        4.3.3.5   *Employment Creation and Income
                Generation*   28

4.4   BIOSOLIDS FROM MUNICIPAL SEWAGE
     SLUDGE   28

    4.4.1   Characteristics of the Biosolids   29

    4.4.2   Benefits of Using Biosolids   31

        4.4.2.1   *Income for the Municipalities*   31

        4.4.2.2   *Closed Nutrient and Carbon Cycle* 31

        4.4.2.3   *Reduced Odors and Flies*   32

CHAPTER 5 ▪ Economic Considerations   33

5.1   NET PRESENT VALUE   33
5.2   INTERNAL RATE OF RETURN   34
5.3   PAYBACK PERIOD   34
5.4   BREAKEVEN POINT   35
5.5   SENSITIVITY ANALYSIS   35
5.6   RISKS ASSOCIATED WITH RESOURCE
     RECOVERY FROM MUNICIPAL PLANTS   35

CHAPTER 6 ▪ Resource Recovery from Chitungwiza,
Firle and Crowborough Plants in
Harare, Zimbabwe: A Case Study   39

6.1   KEY HIGHLIGHTS   39

6.2   INTRODUCTION   41

    6.2.1   Problem Statement   42

    6.2.2   Study Objectives   42

6.3   BACKGROUND   43

    6.3.1   General   43

    6.3.2   Biogas Production Process   43

6.4   GENERAL BIODIGESTER DESIGN FOR
BIOGAS PRODUCTION   44

6.5   RECOVERY OF TREATED EFFLUENT   46

    6.5.1   Introduction   46

    6.5.2   Experimental Approach   46

    6.5.3   Results and Discussion   47

       6.5.3.1   *Raw Sewage and Treated Effluent
Characteristics*   47

       6.5.3.2   *Effect of Hycura Loadings and
Retention Time*   47

       6.5.3.3   *Effect on pH*   49

       6.5.3.4   *Effect on TP*   49

       6.5.3.5   *Effect on TKN*   50

       6.5.3.6   *Effect on BOD*   50

       6.5.3.7   *Effect on COD*   51

       6.5.3.8   *Effect on TSS*   51

       6.5.3.9   *Effect on TDS*   52

       6.5.3.10   *Effect on EC*   52

       6.5.3.11   *Effect on CI⁻ Ion Concentration*   53

6.5.3.12 *Effect on* SO_4^{2-} *Ion Concentration*   53
6.5.3.13 *Effect on DO*   53
6.5.3.14 *Effect on Total* E. coli *and Coliform Content*   53
6.5.4   Summary   54
6.6   BIONUTRIENT RECOVERY   55
6.6.1   Introduction   55
6.6.2   Results and Discussion   55
6.6.2.1   *Bionutrient Removal during Treatment with Hycura*   55
6.6.2.2   *Bionutrient Removal Coefficients*   55
6.6.3   Summary   58
6.7   BIOGAS AND BIOSOLIDS RECOVERY   58
6.7.1   Introduction   58
6.7.2   Experimental Approach   59
6.7.3   Results and Discussion   59
6.7.3.1   *Characterization of the Raw Sewage Sludge*   59
6.7.3.2   *Biogas Production*   59
6.7.4   Summary   64
6.8   PROCESS DESCRIPTION OF CHITUNGWIZA, FIRLE AND CROW BOROUGH MUNICIPAL SEWAGE PLANTS   65
6.8.1   Detailed Process Description for the Municipal Sewage Plants   65
6.8.1.1   *Conventional Plant*   66
6.8.1.2   *Biological Nutrient Removal Plant*   67
6.8.2   Biodigester Section Description   67
6.8.2.1   *Total Solids Content*   68

6.8.2.2	*Temperature*	68
6.8.2.3	*Retention Time*	68
6.8.2.4	*pH*	68
6.8.2.5	*Agitation*	68
6.8.2.6	*Carbon-to-Nitrogen Ratio*	69
6.8.2.7	*Organic (Substrate) Loading Rate*	70
6.8.2.8	*Toxicity*	70
6.8.3	Biogas Purification	70

6.9 MAIN FEATURES OF CHITUNGWIZA, FIRLE AND CROWBOROUGH MUNICIPAL SEWAGE PLANTS 70

6.9.1 Chitungwiza Municipal Sewage Plant 70

6.9.2 Firle Municipal Sewage Plant 71

6.9.3 Crow Borough Municipal Sewage Plant 72

6.10 BIOGAS PRODUCTION IN THE MUNICIPAL SEWAGE PLANTS 73

6.10.1 Possible Amount of Biogas and Electricity Generated at the Three Plants 73

6.10.2 Potential for Biogas Generation in Municipal Sewage Plants 73

6.11 ECONOMIC ASSESSMENT FOR RECOVERING BIOGAS 75

6.11.1 Energy Balance for Electricity Produced in the Three Plants 75

6.11.2 Determination of Economic Feasibility of Biogas Plants 75

6.11.3 Mass, Energy and Economic Balance for Biogas Production at the Chitungwiza, Firle and Crow Borough Municipal Plants 76

CHAPTER 7 ▪ Regulatory Framework and Policy for
Resource Recovery   89

7.1   POLICY PRINCIPLES   89

    7.1.1   Development   89

    7.1.2   Sustainability   89

    7.1.3   Affordability   90

    7.1.4   Accessibility   90

    7.1.5   Gender Equality   90

7.2   NATIONAL RENEWABLE ENERGY POLICY
      PROVISIONS   90

    7.2.1   Poverty Eradication and Employment
            Creation   90

    7.2.2   Promote Local Manufacturing of
            Renewable Energy Technologies   91

CHAPTER 8 ▪ Environmental Impact Assessment   93

8.1   BACKGROUND   95

8.2   SPECIFIC OBJECTIVES TO BE MET   95

8.3   LIKELY SIGNIFICANT ENVIRONMENTAL
      EFFECTS AND THEIR MITIGATION   96

    8.3.1   Site-Specific Factors   96

    8.3.2   Process Technology-Related Effects 96

    8.3.3   Effects Created during Site Preparation
            and Construction   97

    8.3.4   Effects during Operation   97

8.4   MONITORING AND SUPERVISION PROGRAMS   98

    8.4.1   Monitoring   98

    8.4.2   Supervision   98

8.5 SITE PREPARATION AND VEGETATION
CLEARANCE 99

    8.5.1 Impacts 99

    8.5.2 Mitigation 99

8.6 IMPACT: NOISE POLLUTION 99

    8.6.1 Mitigation 100

    8.6.2 Impact: Water Quality 100

    8.6.3 Mitigation 100

8.7 IMPACT: AIR QUALITY 100

8.8 IMPACT: EMPLOYMENT 101

8.9 IMPACT: SOLID WASTE GENERATION 101

    8.9.1 Mitigation 102

8.10 IMPACT: ODOR 102

    8.10.1 Mitigation 102

8.11 STATUTORY INSTRUMENTS 103

CHAPTER 9 ▪ Conclusion and Recommendations 105

9.1 CONCLUSIONS 105

9.2 RECOMMENDATIONS 106

GLOSSARY OF TERMS, 107

REFERENCES, 109

INDEX 117

List of Figures

Figure 1.1   Resource recovery technologies   3

Figure 2.1   Sources of sewage wastewater   6

Figure 3.1   Hycura batch cell kinetics during wastewater treatment. N, number of Hycura cells per given time; 1, lag phase; 2, acceleration phase; 3, decline phase; 4 deceleration phase; 5, stationary phase; 6, death phase   19

Figure 4.1   Biogas production stages   24

Figure 4.2   Biosolids processing from sewage sludge   29

Figure 4.3   Biosolids from sewage treatment process   30

Figure 4.4   Dewatering of biosolids for agricultural land use   31

Figure 6.1   Biogas production stages for electricity generation   44

Figure 6.2   Biogas biodigester schematic diagram   45

Figure 6.3   Bionutrient removal in sewage at a 7-day retention time and Hycura loadings of 0.035 and 0.050 g/L   56

Figure 6.4   Bionutrient removal in sewage at a 40-day retention time and Hycura loadings of 0.035 and 0.050 g/L   56

Figure 6.5   Municipal sewage treatment plant processes   65

Figure 6.6   Effect of pH on methanogen's behavior   69

Figure 6.7   Schematic presentation of the biogas
plants at the three plants   71

Figure 8.1   Environmental impact assessment phases   94

List of Tables

Table 2.1 Municipal Sewage Wastewater Characteristics   7

Table 3.1 Hycura Physicochemical Properties   16

Table 3.2 Potential for Treating Wastewater with Hycura   18

Table 4.1 Properties of Sewage Effluent Used in Irrigation   22

Table 4.2 Municipal Sewage Biogas Composition   23

Table 4.3 Biogas Yield from Various Types of Substrates   25

Table 4.4 Potential Electricity Generation from Methane   25

Table 4.5 Annual Methane Production Rates and Biodigester Sizes for Varying Retention Times   26

Table 4.6 Parameters Affecting Biogas Generation Using Hycura   27

Table 4.7 Municipal Sewage Biosolids Composition   30

Table 5.1 Payback Periods for Biogas Production Reported in the Literature   36

Table 6.1 Sewage Effluent Characteristics   48

Table 6.2 Effect of Hycura on Sewage Wastewater Parameters   54

Table 6.3 Summary of Bionutrient Removal Ratios   58

Table 6.4   Raw Sewage Sludge Characteristics   60

Table 6.5   Biogas Composition from Anaerobic Digestion
of Sewage Sludge   61

Table 6.6   Summary of Biogas Generated in Hycura-
Catalyzed Systems   62

Table 6.7   Sewage Biosolids Quality   63

Table 6.8   Municipal Sewage Capacities for the Three Plants   72

Table 6.9   Municipal Sewage Biogas and Electricity
Production Potential at Chitungwiza, Firle and Crow Borough   74

Table 6.10   Energy Balance for Electricity Generated,
Usage and Potential Surplus at the Three Plants   77

Table 6.11   Chitungwiza Municipal Sewage Plant Mass,
Energy and Economic Balance   78

Table 6.12   Firle Municipal Sewage Plant Mass, Energy
and Economic Balance   81

Table 6.13   Crow Borough Municipal Sewage Plant Mass,
Energy and Economic Balance   84

Table 6.14   Economic Potential for Producing Biogas
from the Municipal Sewage Plants   87

Preface

THIS BOOK PRESENTS AN assessment of the potential to realize financial, economic and social value from municipal plants through resource recovery. In most developing countries, energy and water are scarce commodities. However, large amounts of municipal sewage wastewater are generated daily, and these have the potential to be used as sources for energy, clean effluent and biosolids. Resource recovery using bioaugmented routes from municipal sewage plants promotes sustainability in terms of sewage sludge management as well as wastewater treatment for reuse. Case studies performed for municipal sewage plants indicate the technoeconomic feasibility of recovering a clean effluent, biogas and biosolids, creating a circular economy from municipal plants. Resource recovery from municipal plants provides vast social, environmental and economic benefits to communities.

Authors

Dr. Musaida Mercy Manyuchi is a professional chemical engineer by training and earned a doctorate in technology in chemical engineering from Cape Peninsula University of Technology in South Africa, a master of science in engineering from Stellenbosch University in South Africa and a bachelor of engineering with honors in chemical engineering from the National University of Science and Technology in Zimbabwe. Dr. Manyuchi is a research fellow with the BioEnergy and Environmental Technology Centre in the Faculty of Engineering and the Built Environment at the University of Johannesburg. She is also a faculty member for the Chemical and Processing Engineering Department at the Manicaland State University of Applied Sciences in Zimbabwe. Dr. Manyuchi was a faculty member in the Chemical and Process Systems Engineering Department at the Harare Institute of Technology in Zimbabwe, where she also served as the head of department for 5 years. Dr. Manyuchi was a visiting research scholar at the German Biomass Research Institute in Germany. Dr. Manyuchi is a professional member of the Engineering Council of South Africa, South Africa Institute of Chemical Engineers, Engineering Council of Zimbabwe, Zimbabwe Institute of Engineers and the World Federation of Engineering Organizations.

Dr. Manyuchi's research interest lies in waste-to-energy technologies and the value addition of waste to bioproducts. To date, she has published more than 30 research articles in peer-reviewed journals and conferences, 2 books and 5 book chapters in the waste valorization area. Dr. Manyuchi's work in green initiatives has won several awards, including the prestigious German-based Green Talents, the SanBioFemBiz, the Research Council of Zimbabwe Outstanding Research Award, the Japanese International Award for Young Researchers and Africa Award for AgriTech Innovators.

 **Dr. Charles Mbohwa** is a professor of sustainability engineering and is currently the acting executive dean in the Faculty of Engineering and the Built Environment at the University of Johannesburg. He was the chairman and head of the Department of Mechanical Engineering at the University of Zimbabwe from 1994 to 1997 and was vice-dean of postgraduate studies research and innovation in the Faculty of Engineering and the Built Environment at the University of Johannesburg from 2014 to 2017. He has published more than 350 papers in peer-reviewed journals and conferences, 10 book chapters and 5 books. Upon graduating with his BSc honors in mechanical engineering from the University of Zimbabwe in 1986, he worked as a mechanical engineer with the National Railways of Zimbabwe. Dr. Mbohwa earned a master's in operations management and manufacturing systems from the University of Nottingham. He earned a doctorate in engineering from the Tokyo Metropolitan Institute of Technology in Japan. He was a Fulbright Scholar visiting the Supply Chain and Logistics Institute at the School of Industrial and Systems Engineering, Georgia Institute of Technology; a Japan Foundation fellow; and a fellow of the Zimbabwean Institution of Engineers. He is a registered mechanical engineer with the Engineering Council of Zimbabwe. Dr. Mbohwa has

collaborated with researchers based in many countries, including the United Kingdom, Japan, Germany, France, the United States, Brazil, Sweden, Ghana, Nigeria, Kenya, Tanzania, Malawi, Mauritius, Austria, the Netherlands, Uganda, Namibia and Australia.

 **Dr. Edison Muzenda** is a full professor of chemical and energy engineering and head of the Chemical, Materials and Metallurgical Engineering Department in the Faculty of Engineering and Technology at the Botswana International University of Science and Technology. He is also a visiting full professor of chemical engineering at the University of Johannesburg. He was previously a full professor of chemical engineering and the research and postgraduate coordinator as well as head of the Environmental and Process Systems Engineering and Bioenergy Research Groups at the University of Johannesburg.

Professor Muzenda was also chair of the Process Energy Environment Technology Station Management Committee at the University of Johannesburg. Professor Muzenda earned a BSc with honors (Zimbabwe) and a PhD (Birmingham, United Kingdom). He has more than 20 years' experience in academia, which he gained at various institutions, including the National University of Science and Technology, Zimbabwe; University of Birmingham; University of Witwatersrand; University of South Africa; University of Johannesburg; and Botswana International University of Science and Technology. Through his academic preparation and career, he has held several management and leadership positions. Professor Muzenda's teaching interests and expertise are in unit operations, multistage separation processes, bioenergy and biofuel technologies, waste-to-energy technologies, environmental engineering, chemical engineering thermodynamics, professional engineering skills, research methodology and process economics, management and optimization. He is a recipient of several awards and scholarships for academic excellence, such as the nomination as an outstanding researcher for an African research booklet in 2017 by the Department of Science and Technology, South Africa. His research interests are in bioenergy engineering, sustainable and social engineering, integrated waste management, air pollution and separation processes, as well as phase equilibrium measurement and computation. His current research activities are mainly focused on waste-to-energy projects, particularly biowaste to energy for vehicular application. He has contributed to more than 380 international peer-reviewed and refereed scientific articles in the form of journals, conferences, books and book chapters. He has supervised more than 30 postgraduate students and more than 260 honors and BTech research students. He serves as reviewer for a number of reputable international conferences and journals. Professor Muzenda is a member of several academic and scientific organizations, including the Institute of Chemical Engineers, United Kingdom; South African Institute of Chemical Engineers; and International Society for Development and Sustainability. He is an editor for a number of scientific journals and conferences. He has organized and chaired several international conferences. He currently serves as an editor of the *South African Journal of Chemical Engineering*.

Abbreviations

BNR	Biological nutrient removal
BOD	Biological oxygen demand
CH_4	Methane
COD	Chemical oxygen demand
CO_2	Carbon dioxide
DO	Dissolved oxygen
EC	Electrical conductivity
E. coli	*Escherichia coli*
EMA	Environmental Management Agency
EPA	Environmental Protection Agency
GHG	Greenhouse gas
H_0	Sewage treatment without Hycura added
H_1	Sewage treatment with Hycura added
H_2S	Hydrogen sulfide
IRR	Internal rate of return
kWh	Kilowatt hour
ML/day	Megaliters per day
MPa	Megapascal
MSDS	Material safety data sheet
MW	Megawatt
NPV	Net present value
NTU	Nephelometric turbidity unit
SI	Statutory Instrument
t/day	Tons per day
TDS	Total dissolved solids

TKN	Total Kjeldahl nitrogen
TP	Total phosphates
TS	Total solids
TSS	Total soluble solids
TVS	Total volatile solids
UV	Ultraviolet light
WHO	World Health Organization
ZESA	Zimbabwe Electricity Supply Authority
μS/cm	Microsiemens per centimeter

CHAPTER 1

Resource Recovery from Municipal Plants

A General Review

1.1 BACKGROUND

Water scarcity and energy shortages have become global issues, and there is a need to address wastewater treatment issues using economical treatment technologies that also have the potential to harness energy from municipal plants. Most treatment plants in developing countries still need to identify this potential. A good example of a municipality facing wastewater treatment challenges is Chitungwiza, a satellite town in Mashonaland East in Zimbabwe where sewage is being disposed of without treatment (Nhapi 2009). The disposal of untreated sewage wastewater poses environmental challenges to the receiving water body, Mukuvisi River, which is one of the major rivers providing drinking water to people in Chitungwiza and Harare, Zimbabwe (Nhapi 2009). There is therefore a need to find an economic and alternative method for treating this sewage, concurrently recovering value-added resources, like biogas, which can be converted to electricity, and biosolids, which can be used as fertilizers. Using biological means such as Hycura in enhancing resource recovery strategies in municipal plants that are environmentally friendly and sustainable is a good option for developing countries.

Hycura is an organic biocatalyst bacterium that has been used successfully for treating wastewater in numerous parts of the world, including Australia, Canada and Zimbabwe (Tshuma 2010), as well as enhancing biogas generation through bioaugmentation action. Hycura has been shown to lower river water pollutants based on the biological oxygen demand (BOD), the total dissolved solids (TDS), the chemical oxygen demand (COD), the total Kjeldahl nitrogen (TKN) and the total phosphates (TP) (Tshuma 2010). Hycura has also been reported to enhance bionutrient removal in municipal sewage while recovering biogas and biosolids as value-added products.

1.2 MOTIVATION

Water and energy are scarce commodities globally—hence the need for sustainable treatment of municipal wastewater-recovering energy resources. A good example is the Chitungwiza municipality in Zimbabwe, where poorly treated water comes from Harare City Council, which is located 30 km away (Africa Water Facility 2009; Nhapi 2009), with the satellite town consumption at 30–45 megaliter per day (ML/day) (Nhapi 2009). Water is available to residents only once a week and is priced at $0.38/m^3 (Nhapi 2009), and the biogas being generated from the process is not being recovered, yet the satellite town is also being faced with energy challenges. Due to the water and energy challenges faced by the Chitungwiza municipality in Mashonaland East, Zimbabwe, and all other developing country municipalities, there is a need to sustainably treat municipal wastewater in an integrated manner that allows the recovery of biogas for energy usage and biosolids for use in agriculture.

The untreated sewage effluent from the Chitungwiza sewage plant is disposed into the Mukuvisi River, whereby downstream, the river water is treated for human consumption. This poses problems with the Environmental Management Agency (EMA), which controls water pollution in accordance with Statutory Instrument (SI) 274 of 2004 on effluent and waste disposal regulations (Nhapi and Gijzen 2002). Bionutrient removal in municipal sewage wastewater using viable biocatalysts, while at the same time harnessing value-added biogas and biosolids, makes municipal sewage treatment economically friendly and sustainable.

Using Hycura for possible treatment of municipal sewage wastewater has potential to ease the sewage treatment problems as well as energy problems that are affecting Zimbabwe and other developing countries. Furthermore, recovering biogas may lower the municipal sewage wastewater treatment costs as the biomethane obtained can be used for energy generation at the sewage plant. In addition, the biosolids obtained during the sewage treatment can be used as a biofertilizer, providing an alternative to the chemical industries. Most municipal sewage treatment approaches have

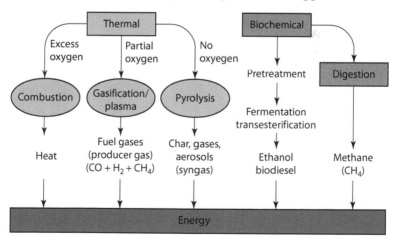

FIGURE 1.1 Resource recovery technologies. (From Davis, R., et al., *Applied Energy*, 88, 3524–3531, 2011.)

been for environmental purposes in order to prevent pollution, preserve the ecosystem, and protect public health. However, an integrated approach to municipal sewage wastewater that focuses on energy recovery will have a significant economic and environmental impact. Apart from anaerobic digestion of the municipal wastewater, other technologies, such as thermal and biochemical conversion routes, can be explored, as shown in Figure 1.1.

CHAPTER 2

Municipal Sewage Wastewater Treatment

2.1 SEWAGE WASTEWATER TREATMENT

Wastewater is any water whose quality parameters have been altered, and it contains a mixture of organic and inorganic pollutants. Wastewater sources include domestic, sewage, municipal, chemical and mining industries, as well as surface runoff and groundwater. Wastewater must be treated for environmental purposes, such as the prevention of groundwater contamination and natural water bodies pollution, preserving the soil, aquatic life and protection of public health (Manyuchi and Phiri 2013; Muserere et al. 2014). Furthermore, wastewater treatment is essential for reuse in agriculture as well as in industry. Sewage wastewater contains pollutants that are physical, chemical or biological in nature. Figure 2.1 shows the various sources of wastewater.

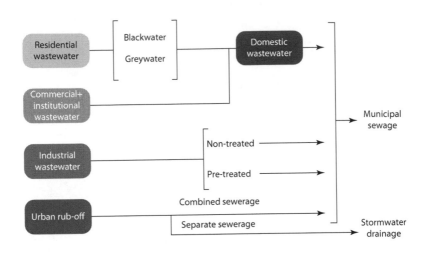

FIGURE 2.1 Sources of sewage wastewater. (From Helmer, R. and Hespanhol, I., Eds., *Water Pollution Control. A Guide to the Use of Water Quality Management Principles*, UNEP, 1999.)

2.2 SEWAGE WASTEWATER CHARACTERISTICS

Sewage is wastewater containing human waste in suspension that is removed from kitchens, lavatories, communities, rainwater flowing into drains and industries; it is mainly composed of 99.9% water and 0.1% fecal matter and urine (Mahmoud et al. 2003; Manyuchi and Phiri 2013; Muserere et al. 2014). Sewage treatment is essential in making sure a clean effluent that can be disposed of to the environment is achieved. The water quality parameters considered in sewage treatment include total fecal coliforms, temperature, dissolved oxygen (DO), biological oxygen demand (BOD), chemical oxygen demand (COD), total phosphates (TP), total Kjeldahl nitrogen (TKN), total suspended solids (TSS), total dissolved solids (TDS) and pH.

An example of raw municipal sewage characteristics from a local sewage plant in Harare, Zimbabwe, is shown in Table 2.1. This is compared with the set standards for acceptable effluent disposal by the Environmental Management Agency (EMA) in Zimbabwe. In order to make sewage disposal safe and environmentally

TABLE 2.1   Municipal Sewage Wastewater Characteristics

Factor	Composition (Muserere et al. 2014)	EMA Effluent Disposal Guidelines (Nhapi and Gijzen 2002)
TSS (mg/L)	250–600	25–50
Fecal content (colonies/100 mL)	200	—
Color	Dark cloudy	—
Temperature (°C)	15–35	25–35
pH	5.5–8.0	6.0–9.0
TKN (mg/L)	20–50	10–20
TP (mg/L)	5–10	0.5–1.5
SO_4^{2-} (mg/L)	50–200	—
CI^- (mg/L)	100–200	—
BOD (mg/L)	100–400	30–50
COD (mg/L)	200–700	60–90

friendly, economic municipal sewage treatment is necessary. Currently, the municipal sewage that is being disposed of, as indicated in Table 2.1, does not meet the required EMA guidelines.

2.2.1 Fecal Coliform Content

The fecal coliform content indicates the bacteria found in the feces of human beings, which contain pathogenic microorganisms. The higher the fecal coliform, the more contaminated the water; therefore, it not useful for human consumption. Recommended fecal coliforms must be less than 200 colonies/100 mL of water sample (Tshuma 2010).

2.2.2 E. coli Content

Escherichia coli is a gram-negative *Bacillus* that belongs to the Enterobateriaceae family and is usually used as a fecal contamination indictor in sewage treatment (Doyle and Padhye 1989). An ideal sample size for *E. coli* content in water treatment will result in 20–80 coliform colonies per dish (Doyle and Padhye 1989).

2.2.3 Temperature

Extremely low temperatures in sewage effluent affect the efficiency of biological treatment systems and the efficiency of sedimentation. The same also applies if the temperatures are on the high side. An acceptable temperature for sewage influent and effluent ranges between 15°C and 35°C (Mahmoud 2002). An increase in the sewage temperature from 15°C to 35°C increases the TSS and COD removal (Ahsan et al. 2005).

2.2.4 Organic Material in Raw Sewage Wastewater

Municipal sewage is mainly composed of organic material. BOD and COD properties are used to determine the oxidation properties of the organic matter in sewage. Sewage generally has a COD/BOD ratio of around 1.7 (Attiogbe et al. 2007). Babaee and Shayegan (2011) used 1.4–2.27 kg/VS (m³·day), and the amount of biogas produced decreased with an increase in organic loading rate.

2.2.4.1 Municipal Sewage BOD$_5$

The BOD is the quantity of oxygen essential for the decomposition of biological organic material in aerobic conditions. The oxygen consumed is related to the amount of decomposable organic matter, and a higher BOD is an indication of high biocontaminants in the wastewater. This has a deadly effect on aquatic life (Muserere et al. 2014). BOD ranges in municipal sewage are between 100 and 400 mg/L (Manyuchi et al. 2013; Muserere et al. 2014).

2.2.4.2 Municipal Sewage COD

The COD is the quantity of oxygen essential for chemical oxidation processes in sewage. The COD does not differentiate between biological oxidizable and nonoxidizable material. The COD values of raw sewage range between 200 and 700 mg/L, and the higher the COD values, the more contaminated the wastewater is (Mahmoud 2002).

2.2.5 Municipal Sewage Total Nitrogen Content

The sources of nitrogen compounds in raw sewage are proteins, amines, amino acids and urea. Ammonia nitrogen results from bacterial decomposition of these organic constituents. Total nitrogen therefore consists of nitrates, nitrites, ammonia and ammonium salts, and these have potential to cause eutrophication if disposed of in water bodies at high concentration (Muserere et al. 2014). Nitrogen is essential for biological protoplasm; hence, it is very important for proper functioning of biological systems. Raw sewage wastewater contains approximately 20–50 mg/L of nitrogen and is measured as TKN (Mahmoud 2002).

2.2.6 Sewage Total Phosphates Content

The sources of TP in raw sewage are mainly food residues containing phosphorous and their breakdowns, which if disposed of in water bodies can result in eutrophication (Muserere et al. 2014). Increased use of synthetic detergents also results in increased phosphorous quantities in sewage. The phosphorous concentration in sewage ranges from 5 to 10 mg/L, and this concentration is adequate to support biological wastewater treatment (Mahmoud 2002).

2.2.7 Sewage pH

The general range for raw sewage is 5.5–8.0 (Muserere et al. 2014). Decomposition of organic matter lowers the sewage pH, while the presence of industrial wastewater may also produce extreme fluctuations. However, pH values of greater than 7.2 are ideal for biogas generation in anaerobic conditions (Mahmoud 2002).

2.2.8 Sewage Color and Odor

Raw sewage has a soapy and cloudy appearance depending on its concentration. As time progresses, sewage becomes stale and darkens in color due to microbial activities (Manyuchi et al. 2013; Muserere et al. 2014).

2.2.9 Sulfate Concentration

Sulfate-reducing bacteria thrive in a pH range of 5–9. The sulfate concentration is generally low in sewage, that is, 50–200 mg/L, and does not affect methanogenic bacteria (Mahmoud 2011). Groundwater infiltration increases the sulfate content, leading to excessive generation of hydrogen sulfide (H_2S). Furthermore, excessive sulfates hinder methanogenic activity (Apte et al. 2011).

2.2.10 Sewage Solids

Raw sewage wastewater contains about 0.1% solids, and these comprise TDS and TSS; although these are present in small quantities, they need proper disposal. Sewage solids comprise dissolved solids, volatile suspended solids and volatile suspended solids. Knowledge on the organic fraction of the solids is necessary as this determines the extent of biological treatment required (Mahmoud 2011).

2.2.11 Chlorides

Chloride concentrations in sewage are greater than the normal chloride content supply. The daily contribution to the chloride concentrations averages 8 g per person. Any abnormal increase in chloride concentration indicates chloride-bearing wastes or saline groundwater infiltration (Apte et al. 2011).

2.2.12 Toxic Metals and Compounds

Toxic metals and compounds such as chromium, copper and cyanide may find their way into municipal sewage wastewater through industrial discharge. The concentration of metals must be monitored to make the biological treatment process efficient, as these have a negative impact on human health (McBride 2003).

2.3 NUTRIENT REMOVAL IN MUNICIPAL SEWAGE USING BIOAUGUMENTATION

The process of municipal sewage management involves the removal of pollutants from the wastewater and includes three major processes: physical methods, chemical methods and biological methods. A safe effluent is normally produced as well as biosolids (Wei et al. 2003; Mahmoud et al. 2003; Tas et al. 2009; Lai et al. 2011; EPA 2013; Palanisamy and Shamsuddin 2013; Lee and Nikraz 2014). The treated sewage effluent has potential to be used for agricultural purposes, such as irrigation water (Manyuchi and Phiri 2013).

Municipal sewage is normally treated in septic tanks, biofilters or aerobic treatment systems. Biological treatment of municipal sewage removes nutrients such as total nitrogen content (TKN), phosphorous content (TP), BOD, COD, TSS, TDS, volatile content, *E. coli*, total coliforms and turbidity (Barker and Dold 1996; Mace and Mata-Alvarez 2002; Ahn et al. 2003; Kraume et al. 2005; Mohan et al. 2005; Kampas et al. 2007).

Municipal sewage wastewater treatment is normally done in three steps, which are the primary, secondary and tertiary stages, like any other wastewater treatment process.

2.3.1 Primary Sewage Treatment

The primary treatment stage consists of temporary holding tanks where heavy solids settle and floating matter is removed from the top. Large objects are also screened from the influent, and a total of about 40%–60% of the suspended solids are removed during this stage. The heavy solids from the primary treatment process have been recommended for use in biodiesel production as they are rich in lipids (Kargbo 2010). Primary treatment of municipal sewage wastewater removes about 25%–35% of sewage suspended solids from the system and about 25%–30% of the sewage BOD (Gupta et al. 2012). The removed solids during primary municipal sewage treatment can also be sent for digestion such that value-added products are obtained.

2.3.2 Secondary Sewage Treatment

The secondary treatment stage involves the removal of dissolved and suspended solids as well as biological matter by biological techniques. This is done by either aerobic or anaerobic bacteria and is normally carried out in activated biosolids systems, trickling filters and rotating biological contactor systems. This is usually done by waterborne microorganisms in a controlled habitat. This stage may also require the separation of microorganisms before tertiary treatment. Secondary sewage treatment removes about 75%–95% of the sewage BOD, and it is at this stage that biogas production occurs for anaerobic treatment (Manyuchi and Phiri 2013). Hycura can be used at this stage as a biological method for the removal of biological contaminants in the wastewater (Dzvene 2013).

2.3.3 Tertiary Sewage Treatment

The tertiary treatment stage is the final stage of sewage treatment. This is done to ensure the safe disposal of the sewage effluent into streams or wetlands or for irrigation purposes. The treated effluent can be disinfected chemically or physically by the use of lagoons or processes such as microfiltration. This stage is essential if the end result of the sewage effluent is to use it for domestic purposes.

2.4 MUNICIPAL SEWAGE TREATMENT CONDITIONS

Municipal sewage treatment can take place either aerobically or anaerobically. Anoxic and anammox conditions can also be considered.

2.4.1 Aerobic Conditions

In aerobic conditions, DO is available and aerobic bacteria survive by metabolizing with the oxygen while at the same time producing carbon dioxide and water as products. The reaction that occurs during aerobic conditions is indicated in Reaction 2.1. The bacteria that facilitate the reaction are called aerobes (Mittal 2011). Besides the presence of oxygen, the rate of the aerobic reaction is also dependent on the retention time, temperature and biological activities of the bacteria (Gupta et al. 2012). Aerobic conditions favor the removal of BOD and COD, dissolved and suspended organics, nitrates and phosphates. However, aerobic treatment of municipal sewage wastewater results in large quantities of biosolids, which may pose problems in terms of disposal. Aerobic treatment of municipal sewage wastewater is normally done in trickling filters, activated biosolids ponds and oxidation ponds (Gupta et al. 2012).

$$\text{Organic Matter} + O_2 + \text{Bacteria}$$
$$\rightarrow \text{New cells} + CO_2 + H_2O + By - products \qquad (2.1)$$

2.4.2 Anaerobic Conditions

During anaerobic conditions, free DO is not available but oxygen is present in the form of sulfate. Anaerobic bacteria can utilize the oxygen bound in the sulfate, but sulfidogenesis must be avoided since this can inhibit the methanogenesis process, especially at a temperature of 30°C and pH of 7–8 (Molwantwa 2002). H_2S, carbon dioxide and biomethane are the major products, as indicated in Reaction 2.2 (Daelman et al. 2012). These gases are collectively termed biogas, and the bacteria that facilitate the anaerobic reactions are called anaerobes (Coppen 2004; Mittal 2011).

$$\text{Organic Matter} + \text{Bacteria}$$
$$\rightarrow \text{Organic acids} + CH_4 + H_2S + H_2O + CO_2 + By - products \qquad (2.2)$$

2.4.3 Anoxic Conditions

In anoxic conditions, DO is not available but oxygen is present in the form of nitrate (NO_3). Facultative bacteria utilize the oxygen bound up in the nitrate for breathing, which results in the release of nitrogen gas (N_2) as a product (Gupta et al. 2012).

2.4.4 Anammox Conditions

An anaerobic ammonium oxidation reaction takes place in the absence of oxygen. During this process, nitrite and ammonium are converted directly to dinitrogen gas and have been reported to have a composition of about 4%–37% loss in nitrogen (Hu et al. 2011). In addition, anammox reactions best occur at 20°C–85°C. Anammox reactions are facilitated by a group of bacteria called planctomycetus, and these include *Brocadia*, *Kuenemia*, *Anammoxglobus* and *Jettenia*. Anammox reactions promote anaerobic COD removal and biogas production while at the same time hindering nitrification and denitrification (Rich et al. 2008).

CHAPTER 3

Anaerobic Treatment of Municipal Wastewater Bioaugmentation

3.1 INTRODUCTION

Several new methods are being applied in municipal wastewater treatment. These include adsorption, disinfection, flocculation, nutrient removal techniques, membrane systems, oxidation, lagoon systems, reed bed technology, vermifiltration technology and water treatment bioaugmentation using Hycura, among others (Zhou and Smith 2002; Coppen 2004; Huang et al. 2010; Tshuma 2010; EPA 2013; Manyuchi and Phiri 2013; Manyuchi et al. 2013, 2015). The anaerobic treatment of municipal wastewater with resource recovery is becoming popular since it provides the plants with stability and income (Lettinga 1995; El-Fadel and Massoud 2001; Mahmoud et al. 2003; Wei et al. 2003; Hospido et al. 2005; Coelho 2006b; USEPA 2013).

The attraction of anaerobic treatment of sewage is the production of biogas, which can be used directly as a source of energy or for electricity generation and has potential to meet the energy requirements of a sewage plant (Lettinga 1995; El-Fadel and Massoud 2001; Mahmoud et al. 2003; Wei et al. 2003; Hospido et al. 2005; Coelho 2006a,b; EPA 2013). The biogas production in sewage plants can be enhanced through bioaugmentation—hence the attraction for using Hycura in enhanced biogas production.

3.2 HYCURA BIOCHEMICAL PROPERTIES

Hycura is a bacteria enzyme that has potential to biodegrade organic waste under both anaerobic and aerobic conditions (Duncan 1970; Powell and Lundy 2007; Tshuma 2010; Dzvene 2013). However, if biogas is to be exploited from the municipal wastewater plants, it is important to consider the anaerobic conditions. During the anaerobic treatment process, Hycura increases in number due to the uptake of nutrients from the wastewater increasing its activity—hence increasing biodegradation rates. Tshuma (2010) reported as much as 2 billion colonies per gram of Hycura added to the wastewater in treatment periods of less than 48 hours. The physicochemical characteristics of Hycura are indicated in Table 3.1.

TABLE 3.1   Hycura Physicochemical Properties

Factor	Explanation
Physical condition	Solid
Color	Light brown
Smell	Faint odor
Boiling point	>100°C
Melting point	Not available
Vapor pressure	Not available
Flash point	Product does not support combustion
Vapor density	Not available
pH	7 when in solution
Solubility	Slightly soluble in cold and hot water

Source: Ecolab, Material safety data sheet, Hycura Compost Accelerator, 903100, 2006, pp. 1–3.

The testing of the microbiological characteristics of Hycura includes catalase, mannitol, urease, Kligler and Indole tests (Schreckenberger and Blazevic 1974; Segal and Potter 2008; Vashist et al. 2013). These have shown the Hycura to be an immotile biocatalyst that contains several enzymes, such as catalase, which has a detoxifying effect; protease, which breaks down proteins in sewage; and amylase, which breaks down the complex sugars to simple sugars in sewage. These characteristics of Hycura promote biogas production in municipal sewage wastewater. Hycura is reported not to promote hydrogen sulfide production, indicating that the biogas produced will have less contaminant and no negative effects, like acid rain, to the environment (Manyuchi et al. 2015).

3.3 USE OF HYCURA IN MUNICIPAL WASTEWATER TREATMENT

The application of Hycura in wastewater treatment results in enhanced biocontaminant removal due to the action of Hycura as it grows and reproduces when inoculated in wastewater during the treatment process. Hycura has been reported to reduce the biological oxygen demand (BOD), total phosphates (TP), total Kjeldahl nitrogen (TKN) and total suspended solids (TSS) by more than 40% when inoculated in wastewater in accordance with studies conducted on swine wastewater, dam water and wool scouring wastewater (Cail et al. 1986; Tshuma 2010; Dzvene 2013). Hycura also has the potential to turn the wastewater pH from acidic to neutral. The removal of biocontaminants in the wastewater has been reported to have a positive effect on the DO, with increases as high as 100% being reported (Tshuma 2010). A summary of the potential for use of Hycura in wastewater treatment for removal of biocontaminants is shown in Table 3.2.

At the end of the treatment process, when all the nutrients in the wastewater are spent, the Hycura dies and becomes part of the biosolids; thus, there is no need for an extra separation process (Tshuma 2010; Dzvene 2013). Hycura can also be used to treat blocked sewer pipes (Tshuma 2010). Hycura also has an odor-eliminating effect, which is a major occurrence in municipal sewage treatment plants, making its usage more attractive (Powell and Lundy 2007). Treatment of wastewater with Hycura results in nitrogen, clean water, micronutrients and biogas (Powell and Lundy 2007). Hycura loading in wastewater ranges between 0.03 and 0.050 g/L at stirring rates of 40–60 rpm (Powell and Lundy 2007).

Municipal sewage wastewater management, including wastewater treatment, is rapidly becoming a critical issue worldwide; therefore, using Hycura in sewage treatment will provide a simple and cost-effective way to treat wastewater (Powell and Lundy 2007).

TABLE 3.2   Potential for Treating Wastewater with Hycura

Parameter	Cail et al. (1986)	Tshuma (2010)	Dzvene (2013)
Type of wastewater	Wool scouring	Dam water	Piggery
BOD (mg/L)	62	96	58
COD (mg/L)	58	—	—
TKN (mg/L)	—	46	36
TP (mg/L)	—	67	—
E. coli (per 100 mL)	—	Positive	—
TSS (mg/L)	—	97	89
TDS (mg/L)	—	—	77
pH	—	6–7	7–8
Temperature (°C)	—	18–24	—
DO (mg/L)	—	100 increase	—
Hycura loading	1% (v/v)	4 kg/day for 1 month in a 5000 m^3 dam (0.024 kg/m^3)	300 m^3 for 150 days (2 m^3/day)
Treatment time (days)	207–211	60	150

COD = chemical oxygen demand; TDS = total dissolved solids; DO = dissolved oxygen.

3.4 HYCURA KINETICS DURING MUNICIPAL WASTEWATER TREATMENT

Hycura has been employed for wastewater treatment in batch systems (Cail et al. 1986; Tshuma 2010; Dzvene 2013). Hycura, like any other cell that is introduced in a batch culture with limited nutrients, goes through various stages of growth in the batch culture; in this case, the nutrients are the biocontaminants in the wastewater. When Hycura is inoculated in wastewater, several growth phases are experienced, and these include the lag phase, exponential phase, deceleration phase, stationary phase, and death phase. The various phases are shown in Figure 3.1.

During the lag phase, the Hycura cells acclimatize to the wastewater treatment environment they are inoculated into. After the acclimatization, they start to grow and reproduce rapidly, resulting in the exponential phase, also known as the logarithmic phase, and it is during this stage whereby the wastewater treatment phase is at its best and the biocontaminant nutrients in the wastewater are utilized. During the exponential phase, the rate of increase in

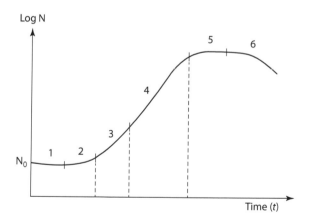

FIGURE 3.1 Hycura batch cell kinetics during wastewater treatment. N, number of Hycura cells per given time; 1, lag phase; 2, acceleration phase; 3, decline phase; 4, deceleration phase; 5, stationary phase; 6, death phase.

the cell number is linear and is dependent on time, as indicated by Equation 3.1:

$$\frac{dX}{dt} = \mu X \qquad (3.1)$$

where:
  μ is the *specific growth rate* of the Hycura cells
  X is the number of Hycura cells per given time

After the exponential phase, the rate of reaction starts to decline, as all the bionutrients would have been depleted and some of the Hycura cells start to die due to lack of food. Afterward, the cells go through the stationary and decline phases. During the death phase, the rate of decline is also exponential, and is represented during the death phase by Equation 3.2:

$$\frac{dX}{dt} = -K_d X \qquad (3.2)$$

where K_d is the death constant for the Hycura cells.

CHAPTER 4

Resource Recovery from Municipal Sewage

4.1 TREATED SEWAGE EFFLUENT

Treating municipal sewage anaerobically results in a clean disposable effluent, biogas and low quantities of biosolids (Lettinga 1995; El-Fadel and Massoud 2001; Mahmoud et al. 2003; Wei et al. 2003; Hospido et al. 2005; Coelho 2006b; USEPA 2013). These are the products that are also promoted by treating wastewater with Hycura (Powell and Lundy 2007). The biosolids produced are in minute quantities as most of them are degraded during the decomposition of the sewage into biogas (El-Fadel and Massoud 2001; USEPA 2013). Anaerobic digestion of sewage has many advantages, which include environmental and economic benefits from the biogas production (Lettinga 1995; El-Fadel and Massoud 2001; Hospido et al. 2005; Coelho 2006a; USEPA 2013). The treated sewage effluent can be used for irrigation purposes or can be disposed of in water bodies depending on the extent of treatment (Manyuchi and Phiri 2013; Muserere et al. 2014). Sewage effluent

TABLE 4.1   Properties of Sewage Effluent Used in Irrigation

Parameter	Irrigation Water	Decorative Water	EMA Disposal Guidelines (Nhapi and Gijzen 2002)
E. coli (per 100 mL)	2	<100	—
BOD (mg/L)	<5	<25	30–50
TKN (mg/L)	—	<10	10–20
TSS (mg/L)	<5	—	25–50
TP (mg/L)	—	<1 mg/L	1–2 mg/L
pH	6–9	6–9	6–9
EC (μS/cm)	<700	—	1000–2000
Turbidity (NTU)	<0.5	—	5
Ammonia (NH$_4$) (mg/L)	<2	—	0.5
CI⁻ (mg/L)	<100	—	—
Na⁺ (mg/L)	<70	—	—

EMA = Environmental Management Agency; TSS = total dissolved solids; EC = electrical conductivity

Source: Al Awadhi, M. A. A. N., Beneficial re-use of treated sewage effluent, Director of Sewage Treatment Plant Department, Dubai Municipality, 2013.

characteristics for use in irrigation and decorative purposes are shown in Table 4.1.

4.2 BIONUTRIENT RECOVERY

The treatment of sewage wastewater with Hycura allows for the recovery of bionutrients, in particular total phosphates (TP) and total Kjeldahl nitrogen (TKN), and sewage bionutrient removal can be measured by its biodegradability and denitrification, which focuses on organic matter removed per sample of sewage (Tas et al. 2009; Lai et al. 2011; Lee and Nikraz 2014). The chemical oxygen demand/biological oxygen demand (COD/BOD), BOD/TKN, COD/TKN and COD/TP ratios are used as good indicators in bionutrient removal through biodegradability and denitrification during biological sewage treatment (Tas et al. 2009). Muserere et al. (2014) reported COD/BOD ratios of 1.5:3.5 and COD/TP ratios of 20:60 for the city of Harare in Zimbabwe water when they were determining its treatability, and the values obtained indicated that the wastewater was easily biodegradable. The recovered nutrients can also be used in the biofertilizer ecosystem.

4.3 BIOGAS FROM SEWAGE SLUDGE

4.3.1 Conditions for Biogas Production

Anaerobic treatment of sewage results in biogas production. Biogas production is optimum at temperature ranges of 35°C–55°C (El-Fadel and Massoud 2001; Wei et al. 2003; Coelho 2006a,b; Kaosol and Sohgrathok 2012). Biogas mainly comprises biomethane, CH_4 (≥60%); carbon dioxide, CO_2 (30%–35%); and traces of H_2S and nitrates (Reali et al. 2001; Arthur and Brew-Hammond 2010; Nazaroff and Alvarez-Cohen 2013; Neczaj et al. 2013). The average biogas composition is indicated in Table 4.2.

The biogas from sewage is generated in a four-step process, which is described below and indicated in Figure 4.1 (El-Fadel and Massoud 2001):

1. Hydrolysis is the breaking down of carbohydrates into simple sugars, proteins into amino acids and fats into fatty acids

2. Acidogenesis involves the decomposition of the lipids, cellulose and proteins into fatty acids using facultative bacteria, for example, *Staphylococcus* into low alcohols and organic acids.

3. Acetogenesis involves the conversion of low alcohols and organic acids into actetic acid, CO_2 and H_2, which are required for the methanogenic process.

TABLE 4.2   Municipal Sewage Biogas Composition

Gas	Composition (%)
CH_4	40–75
CO_2	25–40
N_2	0.5–2.5
O_2	0.1–1
H_2S	0.1–0.5
CO	0.1–0.5
H_2	1–3

Source: Arthur, R. and Brew-Hammond, A., *International Journal of Energy and Environment*, 1 (6), 1009–1016, 2010.

4. Methanogenesis involves the digestion of the fatty acids by methanogenic bacteria, for example, *Methanobacteri* for biomethane production. During this stage, biomethane is produced from the acetogenesis products as well as by-products from the hydrolysis and acidogenesis stages. The equations that occur during this stage are described in Reactions 4.1 through 4.3 (Ozmen and Aslaunzadeh 2009).

$$CH_3COOH \rightarrow CH_4 + CO_2 \qquad (4.1)$$

$$2CH_3CH_2OH + CO_2 \rightarrow CH_4 + 2CH_3COOH \qquad (4.2)$$

$$CO_2 + 4H_2 \rightarrow CH_4 + 2H_2O \qquad (4.3)$$

Biomethane generated from sewage plants can actually meet about 60% of sewer plants' energy requirements (Daelman et al. 2012).

Biogas generated from sewage plants has potential to be applied as a substitute energy resource for heating purposes as well as electricity generation to meet the sewage plants' electricity demands

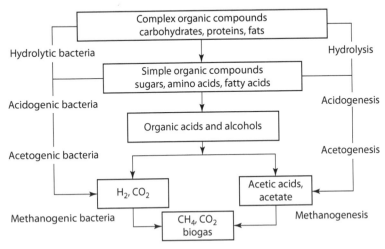

FIGURE 4.1 Biogas production stages. (From Kozani, S. J., Basics of the biogas production process, 2014.)

TABLE 4.3   Biogas Yield from Various Types of Substrates

Substrate	Biogas Yield (L/kg of TVS)
Pig manure	340–350
Vegetable residue	330–360
Sewage biosolids	310–740
Cow manure	90–310

Source: Arthur, R. and Brew-Hammond, A., *International Journal of Energy and Environment*, 1 (6), 1009–1016, 2010.

(Coelho 2006a,b; Malik and Bharti 2009; USEPA 2013; Nazaroff and Alvarez-Cohen 2013). Biomethane produced from sewage plants has high yields of about 310–740 L/kg of total volatile solids (TVS) compared with other sources of organic wastes, such as pig manure, vegetable residue and cow manure (Table 4.3) (Arthur and Brew-Hammond 2010). This indicates that the potential for harnessing biogas from municipal sewage treatment with Hycura is an attractive option (Powell and Lundy 2007).

Methane combustion can result in electricity values of 36.5–37.78 MJ/m^3 (Arthur and Brew-Hammond 2010; Nazaroff and Alvarez-Cohen 2013). Biogas, which was produced from sewage in Ghana with retention times of 10, 20 and 30 days, resulted in 170.72, 341.86 and 419.46 m^3 of methane for sewage plant capacities of 540, 1100 and 1600 m^3 at 30°C, in that order being produced (Arthur and Brew-Hammond 2010). The electricity potential estimation for biogas produced from sewage, which mainly constitutes methane, is indicated in Table 4.4.

TABLE 4.4   Potential Electricity Generation from Methane

Parameter	Value
Methane heating value	37.78 MJ/m^3
Methane content	65%
Biogas engine efficiency	29%
Conversion factor	1 KWh = 3.6 MJ

Source: Arthur, R. and Brew-Hammond, A., *International Journal of Energy and Environment*, 1 (6), 1009–1016, 2010.

TABLE 4.5   Annual Methane Production Rates and Biodigester Sizes for
Varying Retention Times

Retention Time (days)	Annual Methane Estimation (m³)	Annual Energy Production (MWh)	Generator Capacity (kW)	Biogas Digester Size (m³)
10	170.719	6,446,779	50	540
20	341.858	12,915,405	100	1100
30	394.710	14,912,143	120	1600

Source: Arthur, R. and Brew-Hammond, A., *International Journal of Energy and Environment*, 1 (6), 1009–1016, 2010.

A typical biomethane annual production rate and generator capacity for various biogas digesters from sewage are indicated in Table 4.5. High retention times favor increased methane production as well as efficient treatment of the biosolids in terms of pathogenic organisms' reduction (Driessen et al. 2000; Arthur and Brew-Hammond 2010).

4.3.2 Biogas Generation from Municipal Wastewater Using Hycura Bioaugmentation

Biogas has been obtained previously from wastewater treatment using Hycura as a biocatalyst by Cail et al. (1986) from wool scouring wastewater and Duncan (1970) in hog wastes, and high biogas recovery was reported (Table 4.6).

4.3.3 Benefits of Harnessing Biogas from Municipal Plants

4.3.3.1 Alternative and Renewable Energy Source

The recovery of biogas from municipal plants will reduce the dependence on fossil fuels and instead promote the usage of green energy. Biogas from municipal plants is a renewable source of energy. The adoption of biogas as an energy source will promote the preservation of resources and protection of the environment.

4.3.3.2 Reduced Greenhouse Emissions

Using fossil fuels results in the release of carbon dioxide into the environment, unlike when biogas is used. Release of carbon

TABLE 4.6 Parameters Affecting Biogas Generation Using Hycura

Type of Wastewater	Hycura Loading	Retention Time (days)	Organic Loading Rate	T (°C)	pH	Biogas Production Rate	Biomethane Content	Reference
Wool scouring wastewater	1% (w/v)	207–211	0.75–0.99 kg/kg VSS	35°C	7.1–7.4	2.9–3.3 $m^3/(m^3 \cdot day)$; 30% higher compared with a cell free system	68%	Cail et al. (1986)
Hog waste	t0.00625%	50	0.5–1.5 L/day	35°C	7.1–7.2		60% CH_4, 38% CO_2, 1% N_2, 1% water and H_2S traces	Duncan (1970)

dioxide into the environment results in negative effects, such as global warming. Although carbon dioxide is released during the use of biogas, the carbon dioxide generated is assimilated by plants (biomass) during photosynthesis. The replacement of fossil fuels by biogas will result in the decreased release of carbon dioxide and methane, which are the major greenhouse gases in the environment causing global warming.

4.3.3.3 Reduced Dependence on Fossil Fuels

Adopting the recovery and usage of biogas as a renewable energy source will reduce the dependence of most developing countries on imported fuels.

This will in turn increase the national security of local energy supply in substitution for dependence on imports.

4.3.3.4 Waste Reduction and Management

Sewage sludge generated from municipal plants is mostly an uncontrolled waste in most developing countries, and its conversion to biogas will allow for waste reduction and management. Sewage sludge is also a potential methane and carbon dioxide emitter to the environment if not managed properly.

4.3.3.5 Employment Creation and Income Generation

The production of biogas and its recovery from municipal plants need manpower for collecting, production and transportation of materials and equipment. It therefore creates an employment opportunity for the locals, resulting in empowered communities. In addition, the biogas generated from the municipal plants can be sold at a price generating income for the local municipal plants.

4.4 BIOSOLIDS FROM MUNICIPAL SEWAGE SLUDGE

Biosolids are produced during aerobic and anaerobic digestion of sewage, though in smaller quantities by anaerobic processes compared with the aerobic processes (Wei et al. 2003; Mahmoud et al. 2003; Nazaroff and Alvarez-Cohen 2013; USEPA 2013).

Figure 4.2 shows the typical biosolids processing from sewage wastewater.

Biosolids are rich in fertilizer macro- and micronutrients such that they can be utilized as biofertilizers (compost) (Pabsch and Wendland 2013; Nazaroff and Alvarez-Cohen 2013). Figure 4.3 shows typical biosolids from a sewage treatment process.

4.4.1 Characteristics of the Biosolids

The biosolids are stabilized to remove pathogens and offensive odors before use as biofertilizers (Nazaroff and Alvarez-Cohen 2013).

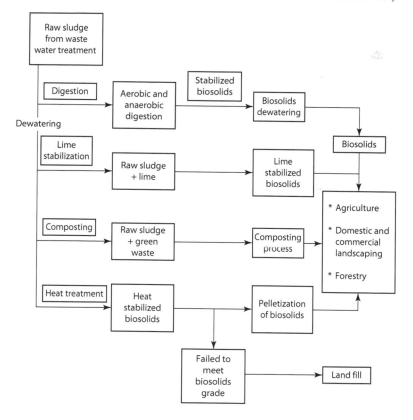

FIGURE 4.2 Biosolids processing from sewage sludge. (From Bharathiraja, B., et al., *Renewable and Sustainable Energy Reviews*, 38, 368–382, 2014.)

FIGURE 4.3 Biosolids from sewage treatment process. (From Nazaroff and Alvarez-Cohen, Anaerobic digestion of wastewater biosolids, Section 6.E.3, 2013.)

Biosolids from wastewater treatment were found to contain nitrogen content of about 2.6%–3.8%, phosphorous content of about 2.0%–2.2% and potassium content of 0.2% (Johannesson 1999; Pabsch and Wendland 2013). The typical biosolids from sewage composition are shown in Table 4.7.

TABLE 4.7   Municipal Sewage Biosolids Composition

Biosolids Parameter	Composition
Total dry solids (%)	2.0–7.0
Volatile solids (%)	60–80
Grease and fats (%)	6.0–30.0
Total nitrogen (%)	1.5–4.0
Phosphorous (%)	0.8–2.8
Potash (%)	0.0–1.0
Cellulose (%)	8.0–15
Iron (%)	2.0–4.0
Silica (%)	15.0–20.0
Alkalinity (mg/L)	500–1500
Organic acids (mg/L)	200–2000

Source: Gorveno, J., et al., Characterization and quantification of Georgia's municipal biosolids production and disposal, University of Georgia, College of Agriculture and Environmental Sciences, Department of Biological and Agricultural Engineering, 2000.

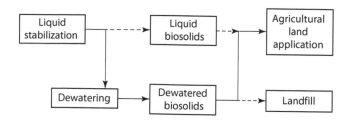

FIGURE 4.4   Dewatering of biosolids for agricultural land use (From CH2MHILL, Niagara region biosolids management master plan, Final report, 2011.)

Biosolids from sewage treatment plants also contain traces of biomethane and pathogens, which are further treated (Daelman et al. 2012). Biosolids processing includes stabilization, dewatering and drying in order to produce biosolids with low moisture content (Pabsch and Wendland 2013). The biosolids dewatering and end-use flow is shown in Figure 4.4.

4.4.2 Benefits of Using Biosolids

4.4.2.1 Income for the Municipalities

The biosolids generated from the anaerobic digestion of sewage can be sold at a cost to the local communities and farmers. This results in additional income generation for the municipalities by having additional income. The biosolids can be used as a biofertilizer due to its high NPK composition.

4.4.2.2 Closed Nutrient and Carbon Cycle

The utilization of biosolids as a biofertilizer provides a closed nutrient cycle as well as a closed carbon cycle. The methane produced by the digestion of sewage is used as a source of bioenergy, while the carbon dioxide is used for photosynthesis. Excess carbon remains in the biosolids and results in enriched soil quality when biosolids are used as a biofertilizer.

4.4.2.3 Reduced Odors and Flies

Biosolids generated from the sewage sludge are odorless, and this reduces the amount of flies in the environment. This creates a sanitary and disease-free environment for farmers and communities.

CHAPTER 5

Economic Considerations

O NE OF THE KEY concerns of resource recovery from municipal sewage plants is the economic benefits of the technology that will be employed. The net present value (NPV), the internal rate of return (IRR), the breakeven point and the payback period are the economic indicators that must be considered when investing in biogas recovery from municipal sewage plants (Gebrezgabher et al. 2010a). In general, economic evaluations of all biogas production scenarios show positive NPV (Gebrezgabher et al. 2010a,b).

5.1 NET PRESENT VALUE

The NPV is normally calculated assuming an even flow of income in accordance with Equation 5.1.

$$NPV = R \times \frac{1 - (1 + i)^{-n}}{1} - \text{Initial investment} \qquad (5.1)$$

where:

R is the net cash inflow expected to be received in each operating plant for the biogas generation period.

i is the expected rate of return in that period.

n is the sewage plant operating period, and normally a maximum of 20 years is given for biogas plants.

5.2 INTERNAL RATE OF RETURN

The investment IRR is defined as the interest rate whereby the NPV for all cash flows either negative or positive from an investment equals zero. The IRR is used as an economic indicator for the attractiveness of an investment. If the IRR surpasses the investment's rate of return, then it must be considered for investment. However, if the IRR is below the anticipated rate of return, then the investment must not be made (Gebrezgabher et al. 2010b). The IRR formula is represented in Equation 5.2.

$$0 = P_0 + \frac{P_1}{(1 + IRR)}$$

$$+ \frac{P_2}{(1 + IRR)_2} + \frac{P_3}{(1 + IRR)_3} \tag{5.2}$$

$$+ \cdots + \frac{P_n}{(1 + IRR)_n}$$

where $P_0, P_1, \ldots, P_n$ are the cash flows for periods 1, 2, ... n, in that order, and IRR is the IRR for the investment.

5.3 PAYBACK PERIOD

The payback period is an economic indicator that is used to evaluate the amount of time required to gain the total initial investment. The calculation of the payback period is represented by Equation 5.3.

$$\text{Payback period} = \frac{\text{Initial investment}}{\text{Periodic cash flow}} \qquad (5.3)$$

Payback periods of less than 5 years have been reported for biogas-producing plants for various sources of sludge, as shown in Table 5.1.

5.4 BREAKEVEN POINT

Breakeven refers to the point of balance between making either a profit or a loss. The breakeven point is calculated by dividing the total fixed expenses by the contribution margin. The contribution margin is sales minus all variable expenses, divided by sales. The formula is represented by Equation 5.4.

$$\text{Breakeven point} = \frac{\text{Total fixed expenses}}{\text{Contribution margin (\%)}} \qquad (5.4)$$

5.5 SENSITIVITY ANALYSIS

Carrying out a sensitivity analysis is important in determining the potential impact for an outcome of a certain economic indicator variable if it is subjected to unforeseen changes (Gebrezgabher et al. 2010a,b). Through the extrapolation and creation of certain scenarios, it can be seen how changes in one economic variable will affect the other variable under study. The variables that can be studied include the payback period, the fixed costs or the production costs.

5.6 RISKS ASSOCIATED WITH RESOURCE RECOVERY FROM MUNICIPAL PLANTS

Resource recovery from municipal plants has its own risks, and these include technical issues, infrastructure and funding or financials. Technical risks may include low biogas generation efficiency and low quality of the biodigester constructed. Infrastructure risk

TABLE 5.1 Payback Periods for Biogas Production Reported in the Literature

Biogas Source	Digestion Period	Biomethane Quality	Electricity Produced	Payback Period	Reference
Cattle dung	49 days	60%	—	0.95 years	Desai et al. (2013)
Human excreta	50 days	—	2710 kWh/day	2 years	Mukumba et al. (2013)
Cow dung					
Chicken manure					

can include the lack of appliances that complement the use of biogas in the municipal plants. Lastly, even though most municipalities in developing countries may want to adopt biogas recovery from municipal plants, they may run into a risk of not having adequate finances for implementation.

CHAPTER 6

Resource Recovery from Chitungwiza, Firle and Crowborough Plants in Harare, Zimbabwe

A Case Study

6.1 KEY HIGHLIGHTS

Treatment of municipal sewage wastewater is a problem in Zimbabwe, yet if the appropriate waste-to-energy technologies are applied, municipal sewage plants can be a source of electrical power instead of relying on power from the national grid, thus minimizing the energy deficit in the country. In this respect, an assessment was conducted for the Chitungwiza, Firle and Crow borough sewage plants' potential to harness biogas for electricity generation with sewage treatment capacities of 19.6, 140 and 54 ML/day, respectively. Plant tours and inspections of the Chitungwiza, Firle and Crow borough plants were conducted, and an understanding of the plant designs as well as the current process flow was attained. Particular emphasis was placed on establishing the availability and state of infrastructure available for the production, handling and storage of biogas. Upon inoculation with Hycura, a clean effluent that meets the Environmental Management Agency (EMA) standards was obtained. Furthermore, the bionutrient removal ratios indicated high biodegradability of contaminants in the municipal sewage wastewater, showing the effectiveness of Hycura. Sewage sludge was collected from the sewage plants and was then placed under conditions that mimic those in a biodigester, and the resultant biogas was collected and analyzed. The biogas was predominantly composed of methane, which had a range of 53%–65%, CO_2 made up 22%–27%, and the balance was trace gases such as H_2S, N_2 and H_2. Experimental results revealed that the use of Hycura as a biocatalyst increases the amount of methane produced to as much as 72%–78%. All three plants have the capacity to produce biogas as they all have biodigesters on site, with 2.65–5.58, 19.47–38.03 and 7.30–21.13 t/day of biomethane being produced from Chitungwiza, Firle and Crow borough plants, respectively. The highest biomethane amounts can be achieved after inoculating 50 g/m^3 of Hycura, a biocatalyst that enriches biomethane production. The Chitungwiza plant, however, does not have the infrastructure to harness the biogas as the digesters are open at the top and there are no biogas holding tanks. Firle and Crow borough plants are equipped with all the basic infrastructure for biogas production, but there is a need for refurbishment of large sections of the plant. Samples of municipal sewage sludge were collected from the different plants and characterized. Using the experimental results together with the design capacities, as well as capacities for the dry and wet seasons of the Chitungwiza, Firle and Crow borough plants, we estimated that the plants have a potential to generate 0.57–1.20, 4.2–8.1 and 1.53–4.56 MW, respectively. An economic assessment indicated the viability of harnessing biogas from the three plants especially after using Hycura as the digestion catalyst. The amount of electricity generated after using Hycura increased by 20%.

6.2 INTRODUCTION

Globally, the energy demand, as well as air and water pollution from the industrial, agricultural and municipal overall operations, continues to rise. There is increased interest in finding out cost-effective and efficient treatment methods for waste generated in the processes. An example of a technology that can be adopted is the anaerobic digestion of sewage sludge from municipal sewage plants using Hycura as the bioaugmentation media. If the municipal sewage plant is optimized, it not only manages wastes but also avails a clean effluent, biogas and biosolids for soil amendment. Recovery of these sources from biogas plants can actually result in sewage treatment plants being profit-making plants that promote sustainable development in developing countries.

Municipal sewage sludge biogas collection and utilization for power generation is also a proven technology to deal with municipal sewage sludge in a sustainable manner. Municipal sewage sludge biogas plants can thus be seriously considered for implementation at the Chitungwiza, Firle and Crow borough municipal sewage plants in Harare, Zimbabwe. This consideration stems from a win-win situation that realizes several environmental benefits and the provision of a renewable source of energy for potential usage at the plant. Methane (CH_4), which is the main component of biogas, is a greenhouse gas (GHG) that has to be mitigated in line with the United Nations Climate Change Conference COP 21 and according to the Kyoto Protocol. The Chitungwiza, Firle and Crow borough municipal sewage plants are currently releasing methane into the atmosphere but have the potential to harness the biogas for usage in the plant, resulting in sustainable development.

The amount of municipal sewage wastewater inflow feeding into the Chitungwiza, Firle and Crow borough municipal sewage plants is rapidly increasing beyond the installed capacity of the plants due to increased population growth and the development of new residential areas merged with a poor standard of living, which also results in an increased generation of waste.

The current environmental, health and energy shortage concerns have resulted in a need to look for alternative sources of energy that utilize the current waste in dealing with the accumulating amounts of sewage sludge in a more economic and sustainable manner. The recovery of the clean effluent, biogas and biosolids in the municipal sewage plants will change the situation by eliminating many negative environmental impacts and also provide a renewable source of energy that can be used to provide power in the plants.

6.2.1 Problem Statement

Currently, municipal sewage plants are not harnessing biogas from sewage sludge, which can be converted to electricity, yet they are using electricity from the national grid, even though the country is currently facing an energy deficit.

6.2.2 Study Objectives

The main objectives of the project were to

1. Evaluate the existing infrastructure for the suitability of biogas production, that is, currently existing biodigesters or ponds

2. Determine the optimal biogas production rates in municipal sewage plants

3. Recommend improvements for biogas harnessing for electricity generation or heating purposes in the plants

4. Conduct a technoeconomic assessment for applying the Hycura technology to optimize sewage sludge conversion to electricity

6.3 BACKGROUND

6.3.1 General

Chitungwiza, Firle and Crow borough municipal sewage wastewater treatment plants are currently meeting their energy demands from electricity from the national grid. The current national energy crisis has resulted in operational difficulties for all these plants. Such a situation has the potential to cause serious disease outbreaks as raw municipal sewage might end up being released into rivers due to the lack of plant availability. The deficit in the conventional energy in Zimbabwe is putting pressure on the economy due to importation of electricity using money that should be used for the socioeconomic development of the country.

The production of biogas, a source of renewable energy, however, offers an alternative that, if managed properly, can help subsidize the amount of conventional energy required. Renewable energy generally refers to energy derived from sources that are not finite. Examples include wind, solar, hydro, geothermal and biomass, which includes biogas from municipal sewage sludge. Utilization of renewable energy promotes sustainable development, economic growth and pollution control techniques. Using biogas from municipal sewage plants will relieve Zimbabwe of the sewage sludge management burden as well as the need to import electricity. The investment in biogas as a form of renewable energy will relieve Zimbabwe's major sewage treatment plants, such as the Chitungwiza, Firle and Crow borough municipal sewage plants, from the burdens of electricity outages and reduce their energy bill. Furthermore, there is potential for the creation of employment opportunities for the surrounding communities, improving their economic status.

6.3.2 Biogas Production Process

The biogas produced from municipal sewage sludge is a mixture of methane, carbon dioxide and hydrogen sulfide. Biogas can be produced under mesophilic (20°C–40°C) and thermophilic (>50°C)

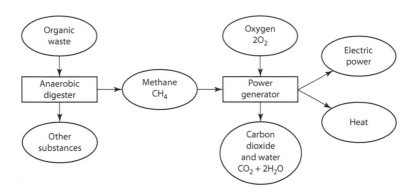

FIGURE 6.1   Biogas production stages for electricity generation.

conditions, and if harnessed, it can be used for cooking, heating and electricity generation purposes (Arthur and Brew-Hammond 2010). The natural anaerobic degradation of municipal sewage sludge in a digester follows a similar pathway to the one that takes place in any anaerobic digester with a net production of biogas (mainly methane). Figure 6.1 depicts the biological transformations that the municipal sewage sludge undergoes under anaerobic conditions, such as hydrolysis, fermentation, acetogenesis and methanogenesis. These biological steps occur concurrently in the biodigester, and microorganisms that are dependent on each other are involved at each stage to enhance the biological process. It is at this stage that Hycura becomes critical as an anaerobic digestion inoculant and the biogas generated will be used for electricity generation. Figure 6.1 shows a summary of biogas production to electricity generation.

6.4 GENERAL BIODIGESTER DESIGN FOR BIOGAS PRODUCTION

Figure 6.2 is an illustration of a basic schematic of a biodigester used on a small scale. This aids in understanding the general biogas production process before delving into the process used in wastewater treatment plants. The design and construction of the biodigester should prevent foreign contamination. Optimal

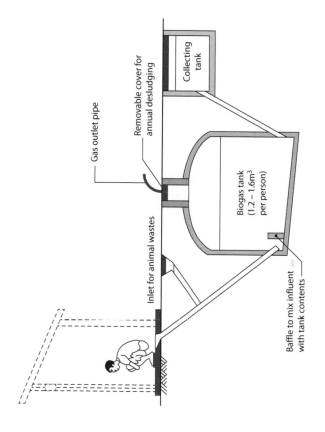

Gas outlet pipe

Removable cover for
annual desludging

Collecting
tank

Biogas tank
(1.2 – 1.6m³
per person)

Inlet for animal wastes

Baffle to mix influent
with tank contents

FIGURE 6.2 Biogas biodigester schematic diagram.

mixing is required in order to fully utilize the organic matter fed into the digester. Once inside the biogas tank, the organic matter undergoes anaerobic digestion like that described in Figure 6.2 under mesophilic conditions. In the biogas collection system, biogas is collected at the top of the digester, while the digestate is collected at the bottom and can be dewatered for use as biofertilizers (Manyuchi et al. 2015).

6.5 RECOVERY OF TREATED EFFLUENT

6.5.1 Introduction

Poorly treated sewage is disposed of in water bodies in developing countries due to poor treatment techniques (Manyuchi and Phiri 2013; Muserere et al. 2014). This poses an environmental threat to the water bodies—hence the need to utilize environmentally friendly treatment techniques. Hycura, an enzyme biocatalyst that possesses biochemical properties that allows it to be used for sewage treatment, can be considered an alternative sewage treatment method (Powell and Lundy 2007). Hycura contains enzymes such as catalase, protease and amylase, which promote detoxification of biological contaminants in sewage, and has the potential to treat wastewater aerobically or anaerobically (Chapter 4). However, there is a need to determine the optimum conditions that can be applied for sewage treatment using Hycura as a biocatalyst, especially the Hycura loadings required and retention time. Hycura can then be applied for sewage treatment and the sewage effluent physicochemical parameters determined so as to ascertain the optimum conditions in sewage treatment as a biological method.

6.5.2 Experimental Approach

Raw sewage was tested for pH, total phosphates (TP), total Kjeldahl nitrogen (TKN), biological oxygen demand (BOD), chemical oxygen demand (COD), total suspended solids (TSS), total dissolved solids (TDS), electrical conductivity (EC), Cl^-, SO_4^{2-} and dissolved oxygen (DO) (Singh and Singh 2010; Popa et al. 2012). The sewage was then treated with Hycura loadings of 0–60 g/m³ between 0 and 70 days (Powell and Lundy 2007). All parameters were in milligrams per liter except for the pH and EC, which was measured in microsiemens per centimeter. Temperature was fixed at 37°C, and the agitation rate in the 500 mL flasks was maintained at 60 rpm. The TKN, TP, BOD, DO and COD ion concentrations were measured using titration methods. TSS and TDS were measured by filtration using the America Public Health Association (APHA) standard methods of determination in wastewater using a 20 μm filter (APHA 2005). The EC and pH were measured by the Hanna HI electrode probe. Lastly, the *Escherichia coli* content was measured through total plate count and the total coliforms by the spread count method. Sewage physicochemical properties were measured for Hycura-free (H_o) sewage effluent and for effluent at varying Hycura loadings and retention time (H_1) for thorough determination of the impact of Hycura on sewage treatment.

6.5.3 Results and Discussion

6.5.3.1 Raw Sewage and Treated Effluent Characteristics

The raw sewage physicochemical characteristics obtained are shown in Table 6.1. All the physicochemical characteristics were off in terms of the EMA guidelines. This made the treatment of sewage in this study using Hycura essential. Although a decrease in the physicochemical properties was achieved for Hycura-free effluent (H_o) (Table 6.1), there was a clear indication that the effluent with Hycura resulted in physicochemical properties acceptable by the EMA (Table 6.1). The decrease in the sewage physicochemical parameters in a Hycura-free system was attributed to the presence of native microbes and enzymes in sewage.

6.5.3.2 Effect of Hycura Loadings and Retention Time

Further work was then done in terms of quantifying the effect of Hycura loading and retention time during municipal sewage treatment. The trends on the sewage effluent physicochemical

TABLE 6.1   Sewage Effluent Characteristics

Parameter	Raw Sewage	Sewage Effluent (H_0)	Sewage Effluent (H_1)	EMA Guidelines
TKN (mg/L)	245 ± 5.5	39.9 ± 0.20	9.4 ± 0.36	10–20
BOD_5 @ 20°C (mg/L)	557 ± 15.3	312.6 ± 0.20	41.8 ± 1.08	30–50
TSS (mg/L)	608 ± 16.1	397.9 ± 0.35	37.3 ± 1.02	25–50
TDS (mg/L)	535 ± 13	253.7 ± 0.25	59.4 ± 0.53	500–1500
E. coli	TMC	TMC	TMC	
EC @ 25°C (μS/cm)	3887 ± 32.1	2186.2 ± 0.21	1070.4 ± 0.36	1000–2000
Cl^- (mg/L)	833 ± 11.2	673.6 ± 0.50	263.3 ± 4.02	–
pH @ 25°C	9 ± 0.3	7.9 ± 0.20	6.3 ± 0.1	6.0–9.0
Coliforms (cfu/mL)	1×10^{11}	1×10^{10}	1×10^{8}	≤1000
TP (mg/L)	52 ± 3.0	29.1 ± 0.20	1.4 ± 0.24	0.5–1.5
SO_4^{2-} (mg/L)	1192 ± 70.8	776.9 ± 0.35	53.6 ± 2.71	–
DO (% saturation)	7 ± 0.2	20.9 ± 0.26	87.0 ± 0.20	≥60
Temperature (°C)	22 ± 1.5	37 ± 0.5	37 ± 0.5	<35
COD (mg/L)	738 ± 12.6	409.5 ± 0.38	77.9 ± 2.24	60–90

H_0 = Hycura-free sewage effluent; H_1 = Sewage effluent with Hycura treatment; TMC = too many to count.

properties were then determined upon increasing Hycura loading and retention time in the digesters.

6.5.3.3 Effect on pH

On treatment of sewage with Hycura, the pH changed from being alkaline to almost neutral. The change in pH varied from 8.3 to 6.3 in the sewage effluent with increased Hycura loading and retention time. An increase in Hycura loading and increase in the retention time in the digester had a positive effect on pH neutralization, and this can be attributed to the removal of contaminants in the effluent due to Hycura activity. Increased Hycura loading increases the rate of biodegradation of biocontaminants as the Hycura will be feeding on the biocontaminants in the sewage, eventually decreasing the pH, especially at longer retention times. pH decreases from 7.4 to 6.8 were observed by Tshuma (2010) for Hycura loadings of 0.024 kg/m³ for a period of 60 days in dam water treatment, indicating an agreement on the impact of Hycura on pH. Dzvene (2013) also reported pH decreases from 7.84 to 7.82 upon adding 2 m³/day of Hycura for 150 days to piggery wastewater, which had a similar trend to observations in this work.

6.5.3.4 Effect on TP

TP exist in sewage as phosphorous. The presence of phosphorus in raw sewage is found both as phosphates and as organically bound phosphorus, and they have a potential to cause eutrophication if uncontrolled. The TP in the sewage effluent decreased by 97% with an increase in Hycura loading, and the retention to around 1.4 mg/L. An increase in Hycura loading and increase in the retention time for Hycura loadings had a positive effect on the TP reduction. The TP reduction is due to the uptake of nutrients in the sewage by Hycura, especially at increased loadings and retention times. A 67% reduction in TP was also reported by Tshuma (2010) after addition of 0.024 kg/m³ of Hycura per day over a 60-day period; however, the reduction in this study is 30% higher because of the higher Hycura loadings reemployed.

6.5.3.5 Effect on TKN

TKN is the measure of the Kjeldahl nitrogen, which is composed of ammonia nitrogen and organically bound nitrogen. TKN, if left uncontrolled, has the potential to cause eutrophication and excessive nitrogen compound release into the environment. Furthermore, the free ammonia part of the total ammonia content is related to an increase in both water pH and temperature for the prevailing conditions in Zimbabwe with water temperatures >20°C. The amount of TKN in the sewage decreased to 9.4 mg/L with an increase in Hycura loading and the retention time. A 96% TKN decrease was achieved for a Hycura loading of 0.035–0.050 g/L at retention times of 7 and 40 days. An increase in Hycura loading and increase in the retention time had a positive effect on the TKN reduction. Tshuma (2010) and Dzvene (2013) also reported a 47% and 48% decrease in TKN for dam water and piggery wastewater, respectively, for a treatment period of 60 and 150 days, respectively, as well as Hycura loadings of 0.024 kg/m^3 and 2 m^3/day, respectively. The TKN reduction in this study is 38% higher than that in the previous two studies. This clearly indicates the importance of using Hycura in bionutrient removal since Hycura utilizes the TKN during its metabolism, especially at higher Hycura loadings and retention times of 40 days.

6.5.3.6 Effect on BOD

High BOD and COD values negatively impact aquatic life in wastewater by lowering the oxygen levels in the wastewater, killing aquatic life. High TKN and TP values can also increase the BOD. The BOD in the sewage decreased by 92% to 41.8 mg/L with an increase in Hycura loading and the retention time in the sewage effluent. Cail et al. (1986), Tshuma (2010) and Dzvene (2013) also observed a 68%, 96% and 58% decrease in BOD upon addition of Hycura to wool scouring wastewater, dam water and piggery wastewater for retention periods of 207, 60 and 150 days, respectively. An almost equal BOD reduction (>90%) in this work was reported by Tshuma (2010) as the retention periods were almost equal. The decrease in BOD was attributed to the increased biodegradation activity of Hycura. However, one observation to note is that at longer retention periods, above 60 days, the Hycura activity decreased. This behavior was possibly due to the degeneration and depletion of nutrients in the wastewater.

6.5.3.7 Effect on COD

COD measures the amount of organic pollutants in sewage. The COD in the sewage decreased linearly by 89.4% to 77.9 mg/L with an increase in Hycura loading and the retention time in the sewage effluent. Only an increase in Hycura loading and the retention time had a positive effect on the COD reduction. The decrease is associated with Hycura activity, which promotes biodegradation in sewage, removing biocontaminants that also contribute to the COD concentration, especially at higher loadings and retention times. Cail et al. (1986) also observed a 58% decrease in COD after treating wool scouring wastewater using Hycura loadings of 1% (v/w) over a period of 207 days. However, this work reported a 30% higher COD reduction than that of Cail et al. (1986) since higher Hycura loadings were used, and optimal removal was achieved at 0.050 g/L and a retention time of 40 days, whereby at this stage the Hycura would not have reached the death phase.

6.5.3.8 Effect on TSS

TSS measure the turbidity of water. High TSS and TDS values, if uncontrolled, can promote the limitation of DO available to aquatic life. The TSS in the sewage decreased by 94% to 37.3 mg/L with an increase in Hycura loading and retention time. An increase in Hycura loading from 0.035 to 0.050 g/L and an increase in the retention time from 7 to 40 days had a positive effect on TSS reduction. The results for TSS reduction in this work are similar to those in the works of Tshuma (2010) and Dzvene (2013). Tshuma (2010) and Dzvene (2013) reported a decrease of at least 88% in dam waster and piggery wastewater upon Hycura loading of 0.024 kg/m^3 and 2 m^3/day, respectively, for 60 and 150 days, respectively. This indicates the ability of Hycura to biodegrade the solids during wastewater treatment.

6.5.3.9 Effect on TDS

TDS refers to the amount of mobile charged ions, including minerals, salts and metals. The TDS in the sewage decreased linearly by 88.9% to 59.4 mg/L with an increase in Hycura loading and the retention time. The 88.9% TDS decrease was due to an increase in Hycura loading and the retention time. The TDS reduction in this study is 12% higher than that in the work of Dzvene (2013). Dzvene (2013) reported a 76.6% decrease in TDS upon Hycura loading of 2 m^3/day in a 4500 m^3 pond to piggery wastewater for 150 days. This indicates the possibility of Hycura's reducing effect on TDS due to its metabolism; however, at very long time retention times, the activity decreases, possibly due to the fact that all the biocontaminants will be used up.

6.5.3.10 Effect on EC

EC refers to the amount of dissolved ions in water that have the potential to conduct electricity. The EC is also directly affected by ions in wastewater, such as chlorides, ammonia and sulfates. The EC in the sewage decreased by 72% to 1070.4 μS/cm with an increase in Hycura loading and the retention time, unlike in a system without Hycura. An increase in Hycura loading from 0.035 to 0.050 g/L and an increase in the retention time in the digester from 7 to 40 days had a positive effect on the EC reduction. EC reduction has a direct relationship to TSS and TDS reduction, which are reduced by Hycura action on the sewage contaminants.

6.5.3.11 Effect on Cl⁻ Ion Concentration

The Cl⁻ ion concentration in the sewage decreased by 68% to 263.3 mg/L in the sewage effluent with an increase in Hycura loading and the retention time. An increase in Hycura loading, an increase in the retention time, and their interaction had a positive effect on the Cl⁻ ion reduction.

6.5.3.12 Effect on SO_4^{2-} Ion Concentration

The SO_4^{2-} ion concentration in the sewage decreased by 92% to 53.6 mg/L in the sewage effluent with an increase in Hycura loading and the retention time, unlike in a system without Hycura (Table 6.1). This decrease can be attributed to the H_2S production hindering activities of Hycura (Cail et al. 1986). This is a very significant increase in Hycura loading and retention time.

6.5.3.13 Effect on DO

The DO concentration in the sewage effluent increased linearly by 222% to 87 mg/L with an increase in Hycura loading and the retention time, unlike in a system without Hycura (Table 6.1). An increase in Hycura loading and an increase in the retention time had a positive effect on the DO increase. This was attributed to the removal of all contaminants due to the Hycura action.

6.5.3.14 Effect on Total E. coli and Coliform Content

The E. coli in the sewage after treatment with Hycura was too high to count. In addition, the total coliform value was too high, that is, around 10^8 coliforms. Despite, the E. coli and total coliforms, the sewage effluent met the prescribed guidelines for sewage disposal in accordance with EMA (Table 6.1). Tshuma (2010) also reported a slight decrease in E. coli composition after addition of $0.024 \, kg/m^3$ Hycura to dam water; however, after 60 days, the dam water still tested positive for E. coli. A summary of the effect of optimal Hycura loading (0.050 g/L) and retention time (40 days) on sewage physicochemical parameters in relation to other wastewater treated with Hycura is given in Table 6.2.

TABLE 6.2   Effect of Hycura on Sewage Wastewater Parameters

Parameter	Cail et al. (1986)	Tshuma (2010)	Dzvene (2013)	This Study
Type of wastewater	Wool scouring	Dam water	Piggery	Sewage
BOD (mg/L)	62.4	96	58.1	92
COD (mg/L)	58.4	—	—	89.4
TKN (mg/L)	—	46	35.6	96
TP (mg/L)	—	67	—	97
Ammonia (mg/L)	—	53	48	—
E. coli	—	Positive	—	Positive
TSS (mg/L)	—	97.6	88.5	94
TDS (mg/L)	—	—	76.6	88.9
pH	—	6.8–7.4	7.82–7.84	6.8–7.4
Temperature (°C)	—	18°C–24°C	—	35°C
DO (mg/L)	—	100 increase	—	222 increase
Hycura loading	1% (v/v)	4 kg/day for 1 month in a 5000 m³ dam (0.024 kg/m³)	300 m³ for 150 days (2 m³/day)	0.550 g/L
Retention time (days)	207–211	60	150	40

6.5.4 Summary

Hycura effectively treats sewage, removing all the wastewater contaminants to meet the required guidelines for effluent disposal compared with treatment without Hycura. Sewage treated with Hycura showed >60% reduction in the sewage contaminants at optimum loadings of 0.050 g/L and a retention time of 40 days. Hycura utilizes the nutrients during its metabolism, making it suitable for biological sewage treatment. However, chlorination must be done to remove the E. coli and the total coliforms that will still be in the treated effluent.

6.6 BIONUTRIENT RECOVERY

6.6.1 Introduction

Excess TKN and TP exist as bionutrients in sewage and can result in eutrophication if disposed of untreated to water bodies, therefore leading to significant destruction of water bodies (Muserere et al. 2014). The biological contaminant removal in sewage can be measured by its biodegradability and denitrification, which focuses on organic matter removed per sample of sewage (Tas et al. 2009; Lai et al. 2011; Lee and Nikraz 2014). The COD/BOD, BOD/TKN, COD/TKN and COD/TP ratios are used as good indicators in bionutrient removal through biodegradability and denitrification (Tas et al. 2009; Muserere et al. 2014). In this chapter, Hycura was used for sewage treatment, and the rate of biodegradability and denitrification of the organic pollutants was measured to determine the bionutrient removal.

6.6.2 Results and Discussion

The raw sewage had a BOD, TKN, COD and TP of 557, 245.2, 52.5 and 739.1 mg/L, respectively. The other raw sewage parameters are given in detail in Table 6.1.

6.6.2.1 Bionutrient Removal during Treatment with Hycura

Sewage treatment with Hycura (A_1 and A_2) resulted in significant bionutrient removal compared with Hycura-free systems (A_0), as shown in Figures 6.3 and 6.4. Significant reduction in TKN, BOD, TP and COD was achieved at Hycura loadings of 0.050 g/L, especially at 40 days' retention time (Figures 6.3 and 6.4). This is attributed to the biocatalytic capability of Hycura in reducing organic pollutants, which becomes more effective with an increase in retention time during treatment.

6.6.2.2 Bionutrient Removal Coefficients

6.6.2.2.1 COD/BOD Ratio   The COD/BOD ratios were all greater than 1.26, which indicated a high rate of biodegradability in

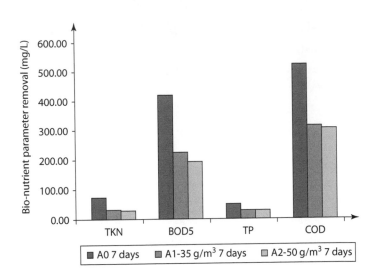

FIGURE 6.3   Bionutrient removal in sewage at a 7-day retention time and Hycura loadings of 0.035 and 0.050 g/L.

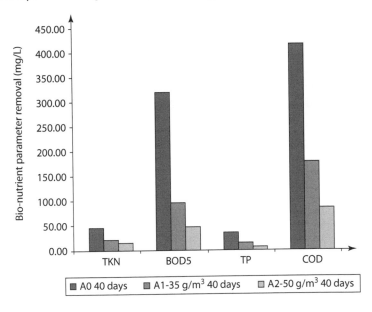

FIGURE 6.4   Bionutrient removal in sewage at a 40-day retention time and Hycura loadings of 0.035 and 0.050 g/L.

sewage and therefore high bionutrient removal, especially at higher retention times and increased Hycura loading. COD/BOD ratios greater than 0.5 are an indication of high degradability of the wastewater (Gomes et al. 2013; Abdalla and Hammam 2014; Muserere et al. 2014). The COD/BOD ratios were greatest at increased Hycura loadings and retention time possibly because Hycura had enough time to acclimatize in the sewage and perform its biocatalytic role.

6.6.2.2.2 BOD/TKN Ratio The BOD/TKN ratios were greater than 4.0 for all the treatment combinations with Hycura at varying retention times. All the BOD/TKN values were greater than 3.0, which was an indication of low biological denitrification in the sewage, at the same time achieving clear effluent at the end (Shin et al. 2005). However, favorable BOD/TKN ratios were achieved at 40 days' retention time. This is an indication that Hycura promotes denitrification in sewage.

6.6.2.2.3 COD/TKN Ratio The COD/TKN ratios were greater than 8.0 for the various Hycura loadings and retention times during sewage treatment. The higher values of the COD/TKN ratio indicated a moderate rate of denitrification in sewage (Shin et al. 2005). However, high denitrification was achieved at increased Hycura loading at the 40-day period. This indicated that Hycura has the potential for denitrification of nitrogen contaminants in sewage promoting biodegradability. Higher retention times are encouraged, as this results in lowered COD/TKN ratios and hence improved denitrification.

6.6.2.2.4 COD/TP Ratio Lastly, the COD/TP ratios were greater than 15.0 for the various Hycura loadings and retention times. COD/TP values of 20–60 were also reported for Harare city wastewater in Zimbabwe by Muserere et al. (2014), and they indicated that values within this range show that the wastewater has

TABLE 6.3   Summary of Bionutrient Removal Ratios

Ratio	Values
COD/BOD	1.26–1.93
BOD/TKN	2.28–10.53
COD/TKN	3.02–16.1
COD/TP	14.07–58.98

increased biodegradability efficiency. TP and nitrogen removal in sewage water is essential in avoiding eutrophication in water bodies, such as rivers and lakes in which it is disposed.

Table 6.3 gives a summary of the bionutrient removal ratios in the wastewater.

6.6.3 Summary

Hycura addition during anaerobic treatment of sewage promotes bionutrient removal through biodegradability and denitrification measured through the COD/BOD, BOD/TKN, COD/TKN and COD/TP ratios. Optimum bionutrient removal was achieved at 0.050 g/L and 40 days' retention time.

6.7 BIOGAS AND BIOSOLIDS RECOVERY

6.7.1 Introduction

Municipal sewage sludge management is increasingly becoming a problem in developing countries due to poor wastewater treatment methodologies. About 60% of the sewage sludge ends up in landfills. Sewage sludge, like any other wastewater sludge, can be used for biogas production using Hycura as a biocatalyst anaerobically (Duncan 1970; Cail et al. 1986). Although biogas has been produced from Hycura-catalyzed wool scouring wastewater sludge by Cail et al. (1986) and from hog wastes by Duncan (1970), the effect of temperature on biogas production under mesophilic and thermophilic conditions still needs to be understood. Furthermore, the quantification of the biosolids that are generated from the digestion process still needs to be understood. Mesophilic conditions are normally employed for wet substrates with TSS of less than or equal to 15% and residence times of 60–95 days, and complete mixing is required, whereas thermophilic conditions are required for wet substrates with TSS greater than or equal to 20% and residence times of 9–45 days (Vindis et al. 2009; Kardas et al. 2011). This chapter therefore focuses on sewage sludge digestion utilizing Hycura focusing on the optimum biogas and biosolids production temperatures.

6.7.2 Experimental Approach

The municipal waste sewage sludge was first dewatered to remove water. Afterward, it was dried to a moisture content of up to 60%. The moisture content and volatile matter content were measured using an AND moisture analyzer as a percentage. Moisture content was measured at 105°C after heating the sample for 30 minutes, while volatile matter was measured after heating for 3 minutes. Sewage sludge digestion was carried out at 37°C under mesophilic conditions in a lab-scale biodigester, which was used to emanate the real situation. Thermophilic conditions of 55°C were also investigated to quantify the effect of temperature.

6.7.3 Results and Discussion

The effect of mesophilic and thermophilic conditions on the various biogas constituents is discussed. Furthermore, the impact of the mesophilic and thermophilic conditions on biosolids generation is also discussed.

6.7.3.1 Characterization of the Raw Sewage Sludge
The sewage had total solids of 1143 mg/L (Table 6.1), and the pH changed from being acidic to alkaline during the digestion process with Hycura under mesophilic conditions. The other raw sewage sludge physicochemical characteristics are shown in Table 6.4.

6.7.3.2 Biogas Production
Biogas production in a digester with substrate activated with Hycura started immediately, resulting in a low-lag phase (Duncan 1970).

TABLE 6.4   Raw Sewage Sludge Characteristics

Parameter	Value
pH	6.3–8.3
COD (mg/L)	750 ± 12.5
TS (mg/L)	1143 ± 14.4
VS (%)	2.5 ± 0.1
Moisture content (%)	60 ± 20
TKN (mg/L)	245 ± 5.1
TP (mg/L)	52.5 ± 2.7
BOD (mg/L)	557 ± 2.5

The biogas obtained had a CH_4 composition of 72%–78%, CO_2 composition of 16%–20%, and trace gases' composition of 8%–12%. Biogas production increased with an increase in Hycura loading from 0 to 0.050 g/L for both the mesophilic (37°C) and thermophilic (55°C) conditions and all sewage sludge loadings. Hycura loadings of 0.050 g/L and sewage sludge loadings of 7.5 g/L·day were found to be optimal in terms of biodegradability of sewage using Hycura. Sewage sludge loadings of 6.9–9.2 g/L·day have been recommended for optimal biogas production (Hesnawi and Mohamed 2013). However, maximum biogas was achieved at mesophilic conditions that were about 50% higher than the thermophilic conditions. This can be attributed to the Hycura activity being optimum at temperatures ~37°C. The biogas produced from Hycura-catalyzed digestion trend was unlike that in other biocatalyst-free municipal sewage digestion reactors whereby the cumulative biogas amount actually increased by more than four times at thermophilic conditions (Vindis et al. 2009; Kardas et al. 2011).

A methane-rich biogas was produced with CH_4 composition ranging from 72%–78%, with a peak being obtained for sewage loadings of 7.5 g/L·day at a Hycura loading of 0.050 g/L compared with Hycura-free digesters, which had a methane composition of 53%–65% (Table 6.5). The biogas also contained 16%–20% CO_2 and trace amounts of 8%–12% H_2S, N_2, and H_2. The trace gases were in lower quantities in digesters with Hycura due to the

TABLE 6.5 Biogas Composition from Anaerobic Digestion
of Sewage Sludge

Gas	% (Hycura)	% (Without Hycura)
CH_4	72–78	53–65
CO_2	16–20	22–27
Traces (H_2S, N_2, H_2)	5–9	8–12

ability of Hycura to hinder their production, effectively improving the quality of the biogas (Table 6.6). The methane was found in high quantities due to the enhanced biodegradability of the sewage sludge by Hycura.

The methane quantity produced in this bioaugmented system was 23% in comparison with previous studies by Duncan (1970), who reported a 60% methane content after digesting hog waste, and by Cail et al. (1986), who reported a 68% methane content after digesting wool wastewater.

6.7.3.2.1 Methane Generation CH_4 production was maximal at mesophilic conditions by more than 100% compared with thermophilic conditions since this is where Hycura activity is at its peak. The CH_4 yield was ~78% for the mesophilic conditions, while it was around 40% for thermophilic conditions at sewage sludge loadings of 7.5 g/L·day. This clearly indicated that mesophilic conditions are favorable for Hycura-catalyzed biogas production from municipal sewage sludge as a value addition strategy.

6.7.3.2.2 Carbon Dioxide Generation   The CO_2 produced was 45% higher in mesophilic conditions than in thermophilic conditions. This was generally attributed to the decreased Hycura activity and other naturally existing microorganisms during the digestion process, resulting in the municipal sewage sludge being partially digestion. However, for both mesophilic and thermophilic conditions, the CO_2 produced significantly decreased with an increase in Hycura loading from 0.035 to 0.050 g/L due to the Hycura

TABLE 6.6   Summary of Biogas Generated in Hycura-Catalyzed Systems

Type of Wastewater	Hycura Loading	Retention Time (days)	Organic Loading Rate	T (°C)	pH	Biogas Production Rate	Biomethane Content	Reference
Wool-scouring wastewater	1% (w/v)	207–211	0.75–0.99 kg/kg VSS	35°C	7.1–7.4	2.9–3.3 m³/(m³·day); 30% higher compared with an A_0 system	68%	Cail et al. (1986)
Hog waste	0.00625%	50	0.5–1.5 L/day	35°C	7.1–7.2		60% CH_4, 38% CO_2, 1% N_2, 1% water and H_2S traces	Duncan (1970)
Sewage sludge	0.050 g/L	40	7.5 g/L·day	37°C	400 mL/day	400 mL/day	72%–78% CH_4, 16%–20% CO_2, and traces (H_2S, N_2, H_2)	Current study

activity, which enhances biomethane production and hinders the production of other gases.

6.7.3.2.3 Trace Gas Generation   Trace gases included trace amounts of H_2S, ammonia and some water. The amount of trace gases produced was 48% more in mesophilic conditions than in thermophilic conditions. During thermophilic conditions, the Hycura becomes inactive—hence, minimal biogas is produced.

A summary of systems whereby Hycura has been used for biogas generation is shown in Table 6.6. For this study, only the optimal conditions for maximum biogas yield are presented under mesophilic conditions. In comparison with studies done earlier by Duncan (1970) and Cail et al. (1986), sewage sludge digestion under mesophilic conditions results in a biogas that is at least 12.8% richer in terms of biomethane content (Table 6.6).

6.7.3.2.4 Biosolids Production   Biosolids are generated as digestate during the Hycura-catalyzed digestion of sewage sludge. These biofertilizers can be utilized as an alternative source of biofertilizers. The biosolids generated had a nitrogen, phosphorous and potassium composition of $8.17 \pm 0.15\%$, $5.84 \pm 0.03\%$ and $1.32 \pm 0.02\%$, respectively (Table 6.7). The biosolids also contained copper $(0.0073 \pm 0.0002\%)$, iron $(0.0087 \pm 0.0003\%)$, calcium

TABLE 6.7   Sewage Biosolids Quality

Parameter	Composition (%)
Nitrogen	8.17 ± 0.15
Phosphorous	5.84 ± 0.03
Potassium	1.32 ± 0.02
Copper	0.0073 ± 0.0002
Iron	0.0087 ± 0.0003
Calcium	0.0079 ± 0.002
Magnesium	0.016 ± 0.0021

(0.0079 ± 0.002%) and magnesium (0.016 ± 0.0021%), which are micronutrients essential for plant growth.

The biosolids obtained by this treatment method can be classified as high-nitrogen-content biosolids (Evanylo 2009). The reduction of water content in the biosolids from 80% to 20%, as well as the hindering effect of Hycura for *E. coli* activity, resulted in a significant decrease in *E. coli* content from 10^{12} to 10^{6} cfu/L. Lower moisture levels and increasing Hycura loading hinder *E. coli* growth, making the biosolids safe for application (Lang et al. 2007; Lang and Smith 2007). The biosolids had a pH of 7.26 ± 0.54, and if a higher pH is needed for application, lime stabilization is recommended.

The amount of biosolids generated was lowest at 5–10 g/L·day at 37°C due to increased digestion from Hycura activity, which was absent at 55°C. As the temperature increased from mesophilic to thermophilic conditions, the amount of biosolids produced showed an exponential decay trend for all the sewage sludge loadings and Hycura loadings. However, if the target's main product is biosolids, thermophilic conditions can be encouraged since they promote pathogen reduction (Willis and Schafer 2006).

6.7.4 Summary

Mesophilic anaerobic digestion of sewage sludge utilizing Hycura at 37°C promotes the production of biomethane-rich biogas, which is almost free of H_2S and nitrogen and has lower CO_2 composition. A Hycura loading of 0.050 g/L and sewage sludge loading of 7.5 g/L·day are essential for optimum biogas production over a 40-day period. Additionally, biosolids that are rich in fertilizer NPK nutrients are produced and can be utilized as biofertilizers. This can result in sustainable management of sewage sludge in developing countries, with high-value-added products like biogas being used for microenergy generation and, at the same time, harnessing biosolids.

6.8 PROCESS DESCRIPTION OF CHITUNGWIZA, FIRLE AND CROW BOROUGH MUNICIPAL SEWAGE PLANTS

6.8.1 Detailed Process Description for the Municipal Sewage Plants

The Chitungwiza, Firle and Crow borough municipal sewage plants all have two different types of plants at their waterworks: the conventional municipal sewage treatment plant and the biological nutrient removal (BNR) plant. Figure 6.5 shows the general flow diagram of both plants. Before municipal sewage is directed to go to the conventional or BNR plants, the raw household and industrial municipal sewage flows to the municipal sewage plants by gravity and then passes through some pretreatment stages. These include the screening stage, which removes floating nonbiodegradable material. Different screen sizes are used, and the screen oversize is removed and either incinerated or sent to the waste dumps. The screening stage also provides uniform, small-sized particles for efficient digestion, as well as protection of the plant

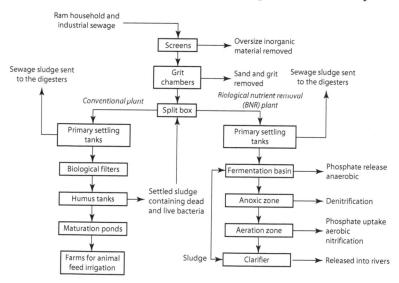

FIGURE 6.5   Municipal sewage treatment plant processes.

equipment from damage. After screening, the municipal sewage passes through a grit chamber for the removal of sand and grits. This is then removed with the aid of airlift pumps. The resultant municipal sewage is then split in a distribution or split box, and the lines and plants to be used are selected. The detailed process description is given below.

6.8.1.1 Conventional Plant

6.8.1.1.1 Primary Settling Tanks   Wastewater that still contains dissolved solids, suspended solids and inorganic constituents is sent to the primary settling tanks, where the suspended solids are allowed to settle by sedimentation or gravity settling. The resulting underflow from the settling tanks is high in solid particles and is known as municipal sewage sludge. This sludge is then pumped to the digester section, where biogas is produced.

6.8.1.1.2 Biological Filters   The overflow from the settling tanks that contain dissolved organic materials and fine suspended particles that do not readily settle is sent to the biological filters. These are essentially large tanks that are filled with rock media. The rock media is covered with thin films of active bacteria, which act on the organic matter as the wastewater trickles down the rock media into receiving launders.

6.8.1.1.3 Humus Tanks   The wastewater that is collected at the bottom of the biological filters is sent to a set of settling tanks known as humus tanks. Here the active and dead bacteria that will have been stripped off the rock media surface settle at the bottom and are then pumped to the split box, where the bacteria precondition the wastewater.

6.8.1.1.4 Maturation Ponds   The overflow from the humus tanks is fed into a set of maturation ponds where further biological processes act on the wastewater until it reaches acceptable levels; then it is then pumped to farms to irrigate animal feed.

6.8.1.2 Biological Nutrient Removal Plant

6.8.1.2.1 Primary Settling Tanks   Wastewater from the split box first passes through primary settling tanks that act in a manner similar to that described for the conventional plant.

6.8.1.2.2 Fermentation Basin   The overflow from the primary settling tanks is fed into the fermentation basin. Here anaerobic conditions are maintained and very gentle mixing is introduced. Phosphate release also takes place at this stage.

6.8.1.2.3 Anoxic Zone   This is where denitrification of nitrogen takes place.

6.8.1.2.4 Aeration Zone   The wastewater is passed through large aerators that bubble air through the water, creating aerobic conditions. Phosphate uptake occurs at this stage, together with nitrification.

6.8.1.2.5 Clarifier   Wastewater is fed to a clarifier, which removes any suspended solids by sedimentation. The underflow sludge is recycled to the fermentation basin, while the overflow is clean enough to be released directly into the river.

6.8.2 Biodigester Section Description

Municipal sewage sludge from the primary settling tanks is sent to the digester section, where it is treated in the manner shown in Figure 6.5. The sludge is distributed into the different digesters, which typically have a volume of 1400 m^3 and made up are of concrete. Anaerobic digestion of sludge takes place in the digesters, with microorganisms acting on the digestible matter and converting it to biogas. Various operating parameters of the digester must be controlled in order to ensure efficient digestion through the enhanced microbial activity for optimum biogas production and these are described in detail below.

6.8.2.1 Total Solids Content

Municipal sewage total solids content can be described as either low, with a TS content of less than 10%; medium, with a TS content of 15%–20%; or high, with a TS content of more than 20%. Firle and Crow borough digesters use the low-solid systems at 10% total solids, while the Chitungwiza digesters use the medium-solid systems.

6.8.2.2 Temperature

The temperature in the biodigester must be kept as constant as possible to allow smooth biogas production. The digesters in Chitungwiza have no temperature control, while those in the Firle and Crow borough plants are maintained at 37°C, which promotes mesophilic conditions. In order to maintain this temperature, the Harare plants make use of heater rooms. These rooms contain boilers that use the biogas generated in the digesters and sent to the storage tanks to heat water to steam. Once the steam is produced, it is passed via a heat exchanger with sludge from the digester and its heat is adjusted to 37°C.

6.8.2.3 Retention Time

The retention time is the total time required for complete digestion of the sewage sludge to occur. The retention time is also dependent on other process parameters, such as temperature and biowaste composition. Retention times of 15–30 days are recommended for biodigester systems under mesophilic conditions.

6.8.2.4 pH

Since biogas production is a biological process, the pH conditions must be closely monitored. pH values between 6.5 and 7.5 are recommended for optimal biogas production (Mes et al. 2003). Figure 6.6 shows the effect of temperature on methanogen activity.

6.8.2.5 Agitation

Agitation is essential in biogas production as it increases the contact between the Hycura and the sewage sludge. Agitation also

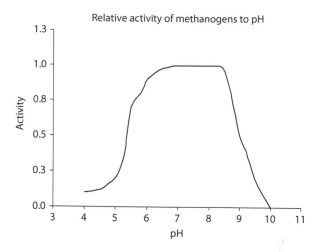

FIGURE 6.6   Effect of pH on methanogen's behavior.

ensures that there is uniform distribution of temperature in the biodigester, lessening the formation of temperature gradients. The agitation rate must also be monitored as there is potential of Hycura cells being destroyed if the agitation rate is high. The mixing system for all the plants that have been looked at is pump mixing. Here sludge is drawn from the bottom of the digester and then fed back from the top. This aids in mixing fresh sludge with matured microorganisms, which can then produce more gas.

Once the biogas has been produced in the primary digesters, it is channeled toward gas holding tanks that store and pressurize the gas where it can then be used either for heating applications, such as heater room applications, or as a source of energy to drive the generation of electricity.

6.8.2.6 Carbon-to-Nitrogen Ratio

The carbon-to-nitrogen (C/N) ratio measures the amount of organic carbon and nitrogen present in the sewage sludge. If the C/N ratio is high, this is an indication that the sewage sludge has low amounts of nitrogen, and this effectively results in low biogas production (Zaher et al. 2007). On the contrary, if the C/N ratio is low, the amount of biogas produced will be high (Zaher et al. 2007). Average C/N ratios of about 24 are acceptable for optimum biogas production.

6.8.2.7 Organic (Substrate) Loading Rate

The organic loading rate of sewage sludge refers to the amount of mass that can be handled per unit volume of the anaerobic digester. The organic loading rate is directly related to the microorganisms' performance in the anaerobic digester, and this also directly affects the amount of biogas generated (Zaher et al. 2007). Higher loadings require higher anaerobic bacteria for the sewage sludge digestion to take place. The sewage sludge loading rate is dependent on parameters such as the digester design and sewage characteristics.

6.8.2.8 Toxicity

Other components in sewage, such as heavy metals like copper and zinc, mineral ions and detergents, can hinder the growth of anaerobic bacteria, and this will eventually affect the amount of biogas produced. High levels of toxicity will result in lower amounts of biogas being produced.

6.8.3 Biogas Purification

Although biogas purification equipment is currently not installed in any of the three plants, it is sometimes important for the biogas produced in the digesters to undergo further purification. The main impurities that are removed are the water vapor, CO_2 and H_2S.

6.9 MAIN FEATURES OF CHITUNGWIZA, FIRLE AND CROWBOROUGH MUNICIPAL SEWAGE PLANTS

The main features of the sewage plants are shown in Figure 6.7.

6.9.1 Chitungwiza Municipal Sewage Plant

Chitungwiza has two plants with a design capacity of 19.6 ML/day; however, currently it has an operating capacity of 23.5 ML/day in the dry season and 33.0 ML/day in the wet season. The BNR plant

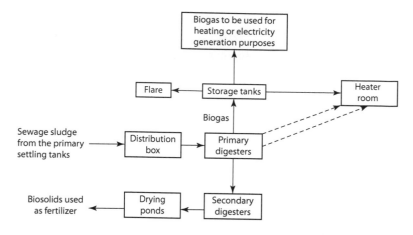

FIGURE 6.7 Schematic presentation of the biogas plants at the three plants.

has two open primary digesters; however, this plant is currently not operational. The conventional plant where municipal sewage is currently being treated has no infrastructure to harness biogas, although municipal sewage sludge is produced.

6.9.2 Firle Municipal Sewage Plant

Firle has a design capacity of 140 ML/day; it is fitted with 19 primary digesters, which have a residence time of 21 days, and three secondary digesters; however, currently the plant is operating at 170–175 ML/day during the dry season and 220–230 ML/day during the wet season. Operation is under mesophilic conditions. The plant is also fitted with a heater room, which consists of boilers that are heated by the biogas produced in the digesters. The heated water is then fed into heat exchangers that maintain the digester temperature at its optimum of 37°C. This temperature is one of the parameters necessary for optimizing biogas generation, together with good mixing and adequate residence time. The plant is also equipped with two biogas holding tanks that store and pressurize the gas. Currently, no biogas is being harnessed despite the infrastructure being available, but Firle has a capacity to produce 97,000 m³ of biogas in 3 hours if the plant is operating normally (Masvingise and Bare 2010). The Firle plant has a capacity to produce 4 MW/day and can save up to $280,000.00 annually if the biogas is harnessed (Kakore 2014).

6.9.3 Crow Borough Municipal Sewage Plant

Crow borough has a design capacity of 54 ML/day, but currently it is operating at 67.5 ML/day in the dry season and 125 ML/day in the wet season. The sludge treatment section consists of a 7×1400 m³ primary digester with a residence time of 21 days under mesophilic conditions and one secondary digester. It also has a heat exchanger room and one biogas holding tank. No biogas is being harnessed at the moment. Information obtained from the assessment indicated the comprehensive features that are necessary for biogas recovery from the sewage plants:

1. Chitungwiza, Firle and Crow borough municipal sewage plants have limited energy resources.

2. The infrastructure for harnessing biogas is available but is not being utilized.

3. The current treatment process being employed does not cater to biogas collection.

The summary of the design and dry and wet season capacities of the plants is given in Table 6.8 based on information found in the plant documents. The current operational capacities are much higher than the set design due to the increased population in Harare.

TABLE 6.8 Municipal Sewage Capacities for the Three Plants

Plant	Design Capacity	Dry Season Capacity	Wet Season Capacity
Chitungwiza	19.6	22–25	28–38
Firle	144	170–175	220–230
Crow borough	54	65–70	120–130

6.10 BIOGAS PRODUCTION IN THE MUNICIPAL SEWAGE PLANTS

6.10.1 Possible Amount of Biogas and Electricity Generated at the Three Plants

The possible amount of biogas to be generated from the three plants is indicated in Table 6.9 for all the various seasons of municipal sewage at the three plants. The electricity production potential was estimated according to Arthur and Brew-Hammond's (2010) conversions and methodology. The values presented hold within a deviation of ±20% of the current values in the sewage plants. Assumptions include

1. The flow rate during the design and dry season has the same solids content, which is about 10% of the sewage, and 0.5% of it is convertible to sewage sludge. The high flow rate is due to increased populations as well as legal and illegal settlements.

2. During the wet season, there is an increase in flow rate due to surface runoff, which is mainly composed of soil and is removed during grit removal during sewage treatment. The remaining sewage has about 10% solids; however, about 0.4% is convertible to sewage sludge due to the dilution by surface runoff.

Municipal sewage consists of approximately 10% municipal sewage sludge, and 90% of it can be converted to biogas during anaerobic digestion and has a density of 260 (kg/m^3).

6.10.2 Potential for Biogas Generation in Municipal Sewage Plants

A similar assessment to other plants in Zimbabwe is possible and recommended because

- There is potential to reduce the energy deficit in the country through harnessing biogas

- Such projects will add to sustainable development as well as socioeconomic enhancement in the country or region

TABLE 6.9 Municipal Sewage Biogas and Electricity Production Potential at Chitungwiza, Firle and Crow Borough

Plant		Flow Rate (ML/day)	Sewage Sludge Quantity (t/day)	Biomethane Produced without Hycura (t/day)	Biomethane Produced with Hycura (t/day)	Electricity Produced without Hycura (MW)	Electricity Produced with Hycura (MW)
Chitungwiza	Design	19.6	101.92	2.65	3.31	0.57	0.71
	Dry season	23.5	122.20	3.18	3.97	0.69	0.86
	Wet season	33.0	137.28	3.57	4.46	0.77	0.96
Firle	Design	144.0	748.0	19.47	24.34	4.20	5.25
	Dry season	172.0	897.0	23.32	29.15	5.03	6.29
	Wet season	225.0	936.0	24.34	30.42	5.25	6.57
Crow borough	Design	54.0	280.80	7.30	9.13	1.53	1.97
	Dry season	67.5	351.00	9.13	11.41	1.97	2.46
	Wet season	125.0	520.0	13.52	16.90	2.92	3.65

- There is expected economic feasibility of such projects

- A high fraction of municipal sewage sludge is generated in the Chitungwiza, Firle and Crow borough municipal sewage plants. The average density of sewage waste in the Chitungwiza, Firle and Crow borough municipal sewage plants is around 1000 kg/m³, has an average caloric value of 10 MJ/kg and rises to about 37 MJ/kg in biogas.

6.11 ECONOMIC ASSESSMENT FOR RECOVERING BIOGAS

6.11.1 Energy Balance for Electricity Produced in the Three Plants

A detailed energy balance was conducted so as to determine the amount of electricity that can be generated from the Chitungwiza, Firle and Crow borough plants. From the assessment indicated in Table 6.10, it is possible to generate biomethane than can sufficiently power the sewage plants, especially systems that are inoculated with Hycura. Excess electricity can also be generated, and this can be sold to the Zimbabwe Electricity Supply Authority (ZESA) so that the energy deficit in the country is minimized.

6.11.2 Determination of Economic Feasibility of Biogas Plants

The economic feasibility of producing electricity from municipal sewage sludge methane can be roughly assessed by the three following criteria according to the U.S. Environmental Protection Agency (USEPA 2012):

1. Municipal sewage plants produce 28,316.85 m³ of CH_4/day.

2. Smaller municipal sewage plants can also be viable depending on the circumstances and solutions available. In particular, municipal sewage sludge gases can be refined for reuse at the municipal sewage plants.

3. Municipal sewage plants with as little as 50,000 tons in place or gas flows of 566.34 CH_4 kg/day can be feasible.

So from the EPA standards and the assessment done from this work, one can see the feasibility of harnessing biogas for energy generation in sewage plants.

6.11.3 Mass, Energy and Economic Balance for Biogas Production at the Chitungwiza, Firle and Crow Borough Municipal Plants

Production of electricity from biomethane is a feasible process, especially when Hycura is incorporated as the digestion catalyst. For all three plants, an average 20% increase was obtained for Hycura-incorporated digestion in terms of the electricity generated (Tables 6.11 through 6.13). This shows the economic viability of generating biogas for reuse in municipal sewage plants, resulting in sustainable development. The following was considered in conducting the economic balance (Table 6.14):

- 60% on average is generated in sewage plants without biocatalysis.

- 75% on average is generated in sewage plants with the aid of Hycura as a biocatalyst.

- The internal rate of return was calculated over a 15-year sewage biogas harnessing period.

- The economic assessment was based on refurbishment of the digestion section only.

TABLE 6.10 Energy Balance for Electricity Generated, Usage and Potential Surplus at the Three Plants

Plant	Flow Rate (ML/day)		Electricity Produced without Hycura (kWh per annum)	Electricity Produced with Hycura (kWh per annum)	Electricity Required in Plant (kWh per annum)	Surplus to the Grid (without Hycura)	Surplus to the Grid (with Hycura)
Chitungwiza	Design	19.6	5,010,203.74	6,262,754.68	4,905,600	104,603.74	135,715.68
	Dry season	23.5	6,007,132.04	7,508,915.05	4,905,600	1,101,532.04	2,603,315.05
	Wet season	33.0	6,748,437.70	8,435,547.12	4,905,600	1,842,837.70	3,529,947.12
Firle	Design	144.0	36,809,660.16	46,012,075.20	36,441,600	368,060.16	9,570,475.20
	Dry season	172.5	44,094,905.40	55,118,631.75	36,441,600	7,653,305.40	18,677,031.75
	Wet season	225.0	46,012,075.20	57,515,094.00	36,441,600	9,570,475.20	21,073,494.00
Crow borough	Design	54.0	13,803,622.56	17,254,528.20	13,665,600	138,022.50	3,588,928.20
	Dry season	67.5	17,250,528.20	21,568,160.25	13,665,600	3,588,928.20	7,902,560.25
	Wet season	125.0	25,562,264.00	31,952,830.00	13,665,600	11,896,664.00	18,287,230.00

TABLE 6.11 Chitungwiza Municipal Sewage Plant Mass, Energy and
Economic Balance

Mass in the Biodigester		
Plant capacity	19.60	ML/day
Municipal sewage sludge 0.3–0.5 of treated wastewater volume	0.01	
Volume of sludge per day	98.00	m³/day
Density of sludge	1.06	Tonnes/day
Total mass of sludge	103.88	Tonnes/day
Mass Out of the Digester		
Assume that 30%–40% of the sewage sludge is organic matter	0.40	
Mass of dry matter	41.55	Tonnes/day
Assume that 50% of the organic dry matter is carbon	0.50	
Mass of carbon content	20.78	Tonnes/day
Assume that 45% of the sewage sludge is digestible	0.50	
Carbon converted to biogas (CH_4, CO_2)	10.39	Tonnes/day
	865.67	kmol/day
Assume that the minimum composition of the biogas is 60%	0.60	
Mass flow rate of CH_4	519.40	kmol/day
	8.31	Tonnes/day
Assume that 37% of the biogas is carbon dioxide	0.37	
Weight of biogas = Weight of CH_4 + Weight of CO_2	320.30	kmol/day
	14.09	Tonnes/day
	22.40	Tonnes/day
Since 50% carbon is digested, 50% is residual organic solids	0.50	
Weight of residual organic solids	10.39	Tonnes/day
Since 65% sludge is organic dry matter, 35% sludge is nondegradable solids	0.35	
Total weight of digestate	36.36	Tonnes/day
	46.75	Tonnes/day

(*Continued*)

TABLE 6.11 (CONTINUED)   Chitungwiza Municipal Sewage Plant Mass, Energy and Economic Balance

Energy Generation: Cogeneration

1 kmol CH_4 = 22.4 m³	22.40	m³
Volume of CH_4	11,634.56	m³/day
1 m³ methane = 36 MJ = 10 kWh	10.00	kWh
Total energy produced by methane per day	116,345.60	kWh/day
Assume that the generator is 35% efficient	0.35	
Reusable	40,720.96	kWh/day
The following efficiencies will be used		
Electric efficiency is 37%	0.37	
Thermal efficiency is 63%	0.63	
(Electrical energy)E_{gen}	15,066.76	kWh/day
	0.63	MW
Amount of heat generated		
(Heat energy)H_{gen}	25,654.20	kWh/day

Economic balance

Biogas plant, including all essential installations, is US$500 per m³ digester	500	
Size of digester	1,400	
Cost of rehabilitation to be 60% of a new biogas plant cost	0.6	
Number of digesters	2	

Capital investment

Item	Equipment	Cost ($)
Biogas plant rehabilitation	Pumps, motors, gearboxes, pipes, instrumentation	840,000.00
Gas purification equipment + installation	Condenser, compressor, piping, instrumentation	800,000.00
Cogenerator equipment + installation	Generator plus supporting equipment	600,000.00
Total capital investment		2,240,000.00

(*Continued*)

TABLE 6.11 (CONTINUED)   Chitungwiza Municipal Sewage Plant Mass, Energy and Economic Balance

Operating costs

Item	Approximations	Amount ($)
Cost of maintenance	5% of investment cost	9,333.33
Operating labor	$400 per person per month × 24	9,600.00
Plant overhead	50% labor costs	4,800.00
Laboratory costs	15% of operating labor	1,440.00
Total cost per month		25,173.33
Total cost per year		302,080.00

Note: Raw material is available free of charge

Revenue

Assume the plant runs for	8,000	h/year
Amount of electricity produced per year	5,022,251.73	kWh
Selling price of electricity per kWh	0.10	$/kWh
Gross income	502,225.17	$
Production cost per year	302,080.00	$
Net profit	200,145.17	
Payback period	11.19	Years

TABLE 6.12   Firle Municipal Sewage Plant Mass, Energy and Economic
Balance

Mass in the Biodigester		
Plant capacity	140	ML/day
Municipal sewage sludge 0.3%–0.5% of treated wastewater volume	0.005	
Volume of sludge per day	700	m³/day
Density of sludge	1.06	Tonnes/day
Total mass of sludge	742	Tonnes/day
Mass Out of the Digester		
Assume that 40% of the sludge is organic matter	0.4	
Mass of dry matter	296.8	Tonnes/day
Assume that 50% of the organic matter is carbon	0.5	
Mass of carbon content	148.4	Tonnes/day
Assume that 50% carbon is biodegraded since the range of biogas production is 30%–60%	0.5	
Carbon converted to biogas (CH_4, CO_2)	74.2	Tonnes/day
	6,183.33	kmol/day
Assume that 60% volume of biogas is methane since the range is 50%–75%	0.6	
Mass flow rate of CH_4	3,710	kmol/day
	59.36	Tonnes/day
Assume that 37% volume of biogas is carbon dioxide	0.37	
Weight of biogas = Weight of CO_2 + Weight of biogas	2,287.83	kmol/day
	100.66	Tonnes/day
	160.02	Tonnes/day
Since 50% carbon is digested, 50% is residual organic solids	0.5	
Weight of digestate	74.2	Tonnes/day
Since 65% sludge is organic dry matter, 35% sludge is nondegradable solids	0.35	
Total weight of digestate	259.7	Tonnes/day
	333.9	Tonnes/day

(*Continued*)

TABLE 6.12 (CONTINUED)   Firle Municipal Sewage Plant Mass, Energy and Economic Balance

Energy Generation: Cogeneration

1 kmol CH_4 = 22.4 m^3	22.4	m^3
Volume of CH_4	83,104	m^3/day
1 m^3 methane = 36 MJ = 10 kWh	10	kWh
Total energy produced by methane per day	831,040	kWh/day
Assume that the efficiency of the generator is 35%	0.35	
Reusable	290,864	kWh/day
The following efficiencies will be used		
Electric efficiency 37%	0.37	
Thermal efficiency 63%	0.63	
(Electrical energy)E_{gen}	107,619.68	kWh/day
	4.48	MW
Amount of heat generated		
(Heat energy)H_{gen}	183,244.32	kWh/day

Economic Balance

Biogas plant, including all essential installations, is US$500 per m^3 digester	500	
Size of digester	1,400	
Cost of rehabilitation to be 60% of a new biogas plant cost	0.6	

Capital Investment

Item	Equipment	Cost ($)
Biogas plant rehabilitation	Pumps, motors, gearboxes, pipes, instrumentation	840,000.00
Gas purification equipment + installation	Condenser, compressor, piping, instrumentation	800,000.00
Cogenerator equipment + installation	Generator plus supporting equipment	600,000.00
Total capital investment		**2,240,000.00**

(*Continued*)

TABLE 6.12 (CONTINUED)   Firle Municipal Sewage Plant Mass, Energy and Economic Balance

Operating costs		
Item	Approximations	Amount ($)
Cost of maintenance	5% of investment cost	9,333.33
Operating labor	$400 per person per month × 24	9,600
Plant overhead	50% labor costs	4,800
Laboratory costs	15% of operating labor	1,440
Total cost per month		25,173.33
Total cost per year		302,080
Raw material is available free of charge		
Revenue		
Assume the plant runs for	8,000	h/year
Amount of electricity produced per year	35,873,226.67	KWh
Selling price of electricity per kWh	0.10	$/kWh
Gross income	3,587,322.67	$
Production cost per year	302,080.00	$
Net profit	3,285,242.67	
Payback period	0.68	Years

TABLE 6.13   Crow Borough Municipal Sewage Plant Mass, Energy and Economic Balance

Mass in the Biodigester		
Plant capacity	54	ML/day
Municipal sewage sludge 0.3%–0.5% of treated wastewater volume	0.005	
Volume of sludge per day	270	m³/day
Density of sludge	1.06	Tonnes/day
Total mass of sludge	286.2	Tonnes/day
Mass Out of the Digester		
Assume that 40% of sludge is organic dry matter	0.4	
Mass of dry matter	114.48	Tonnes/day
Assume that 50% of organic dry matter is carbon	0.5	
Mass of carbon content	57.24	Tonnes/day
Assume that 50% carbon is biodegraded since the range of biogas production is 30%–60%	0.5	
Carbon converted to biogas (CH_4, CO_2)	28.62	Tonnes/day
	2385	kmol/day
Assume that 60% volume of biogas is methane since the range is 50%–75%	0.6	
Mass flow rate of CH_4	1,431	kmol/day
	22.896	Tonnes/day
Assume that 37% volume of biogas is carbon dioxide	0.37	
Weight of biogas = Weight CH_4 + Weight CO_2	882.45	kmol/day
	38.83	Tonnes/day
	61.72	Tonnes/day
Since 50% carbon is digested, 50% is residual organic solids	0.5	
Mass of organic solids	28.62	Tonnes/day
Since 65% sludge is organic dry matter, 35% sludge is nondegradable solids	0.35	
Weight of digestate	100.17	Tonnes/day
	128.79	Tonnes/day

(*Continued*)

TABLE 6.13 (CONTINUED) Crow Borough Municipal Sewage Plant Mass, Energy and Economic Balance

Energy generation: Cogeneration		
1 kmol CH_4 = 22.4 m^3	22.4	m^3
Volume of CH_4	32,054.4	m^3/day
1 m^3 methane = 36 MJ = 10 kWh	10	kWh
Total energy produced by methane per day	320,544	kWh/day
Assume that the efficiency of the generator is 35%	0.35	
Reusable	112,190.4	kWh/day
The following efficiencies will be used		
Electric efficiency is 37%	0.37	
Thermal efficiency is 63%	0.63	
(Electrical energy)E_{gen}	41,510.45	kWh/day
	1.73	MW
Amount of heat generated		
(Heat energy)H_{gen}	70,679.95	kWh/day

ECONOMIC BALANCE

Biogas plant, including all essential installations, is US$500 per m^3 digester		500
Size of digester		1,400
Cost of rehabilitation to be 60% of a new biogas plant cost		0.6

Capital investment

Item	Equipment	Cost ($)
Biogas plant rehabilitation	Pumps, motors, gearboxes, pipes, instrumentation	840,000.00
Gas purification equipment + installation	Condenser, compressor, piping, instrumentation	800,000.00
Cogenerator equipment + installation	Generator plus supporting equipment	600,000.00
Total capital investment		2,240,000.00

(*Continued*)

TABLE 6.13 (CONTINUED)   Crow Borough Municipal Sewage Plant Mass, Energy and Economic Balance

Operating costs

Item	Approximations	Amount ($)
Cost of maintenance	5% of investment cost	9,333.33333
Operating labor	$400 per person per month × 24	9,600
Plant overhead	50% labor costs	4,800
Laboratory costs	15% of operating labor	1,440
Total cost per month		**25,173.33**
Total cost per year		**302,080**

Raw material is available free of charge

Revenue

Assume the plant runs for	8,000	h/year
Amount of electricity produced per year	13,836,816.00	kWh
Selling price of electricity per kWh	0.10	$/kWh
Gross income	1,383,681.60	$
Production cost per year	302,080.00	$
Net profit	1,081,601.60	
Payback period	2.07	Years

TABLE 6.14 Economic Potential for Producing Biogas from the Municipal Sewage Plants

Plant		Flow Rate (ML/day)	Sewage Sludge Quantity (t/day)	Biomethane Produced without Hycura (t/day)	Biomethane Produced with Hycura (t/day)	Electricity Produced without Hycura (MW)	Electricity Produced with Hycura (MW)	Capital Investment without Hycura (US$)	Capital Investment with Hycura (US$)	Payback Period without Hycura	Payback Period with Hycura	Net Present Value without Hycura ($)	Net Present Value with Hycura ($)	Internal Rate of Return without Hycura (%)	Internal Rate of Return with Hycura (%)
Chitungwiza	Design	19.6	101.92	2.65	3.31	0.57	0.71	3,355,882.13	3,641,852.67	10.30	8.19	1,587,009.34	2,166,049.64	5	9
	Dry season	23.5	122.20	3.18	3.97	0.69	0.86	3,583,491.33	3,926,364.17	8.51	6.97	2,050,304.30	2,074,561.80	8	12
	Wet season	33.0	137.28	3.59	4.46	0.77	0.96	3,752,739.20	4,137,924.00	7.63	6.35	2,394,805.68	3,174,737.52	10	13
Firle	Design	144.0	748.0	19.47	24.34	4.20	5.25	23,468,032.00	25,869,040.00	7.56	6.51	15,113,253.84	19,338,210.02	10.0	13.0
	Dry season	172.0	897.0	23.32	29.15	5.03	6.29	25,131,330.00	27,948,162.50	6.62	5.78	1,849,870.85	23,565,798.91	13	15
	Wet season	225.0	936.00	24.34	30.42	5.25	6.57	25,569,040.00	28,495,300.00	6.42	5.62	19,389,822.70	24,678,322.30	13	16
Crow borough	Design	54.0	280.80	7.30	9.13	1.53	1.97	7781 512.00	8,569,390.00	6.97	5.97	5,438,015.59	7,033,033.72	12	15
	Dry season	67.5	351.00	9.13	11.41	1.97	2.46	8,569,390.00	9,554,237.50	5.93	5.15	7 041 728.91	9 035 872.82	15	18
	Wet season	125.0	520.00	13.52	16.90	2.92	3.65	10,466,133.33	11,925,166.67	4.67	4.19	10,902,520.25	13,856,807.51	20	23

Regulatory Framework and Policy for Resource Recovery

7.1 POLICY PRINCIPLES

The renewable energy (RE) policy for Zimbabwe is developed keeping in mind the following key principles: development, sustainability, affordability, accessibility and gender equality.

7.1.1 Development

The RE policy is based on the fact that economic and social development is the basic right of every citizen. The policy aims to achieve overall development of the energy sector in the country, resulting in the economic and social empowerment of the citizens.

7.1.2 Sustainability

People are entitled to a healthy and productive life. However, this should be in harmony with nature. Development today must not undermine the development and environmental needs of future generations. The RE policy aims to develop sustainable energy resources for the country, which will not only benefit the current generation but also create a sustainable future.

7.1.3 Affordability

The RE policy aims to create a balance between project viability and affordable energy cost to provide the citizens of Zimbabwe with cleaner, greener and cheaper energy options.

7.1.4 Accessibility

This key principle in framing this policy is driven by the United Nations SE4ALL program, which aims to reduce the carbon intensity of energy as well as increasing energy access to make it available to everyone.

7.1.5 Gender Equality

The advancement of gender equality and equity is a critical theme that runs through all developmental policies and frameworks in Zimbabwe. The key message here is that the government seeks to achieve a gender-just society where men and women enjoy equality and equity and participate as equal partners in the development process of the country.

7.2 NATIONAL RENEWABLE ENERGY POLICY PROVISIONS

7.2.1 Poverty Eradication and Employment Creation

Poverty eradication is the application of a set of measures, both economic and humanitarian, that are intended to permanently lift people out of poverty.

The measures also aim to remove social and legal barriers to income growth among the poor. The availability of energy in rural communities will unlock their productive potential. There is a significant energy divide between the rich and the poor, between men and women. Access to energy has a significant economic and social impact that varies based on gender, social strata and geographic and demographic segmentation.

This policy promotes the development of RE resources to create opportunities for better health, employment and income generation equally for men and women.

7.2.2 Promote Local Manufacturing of Renewable Energy Technologies

Local manufacturing of RE systems and devices will not only spur growth in the energy sector but also ensure a robust supply chain and generate employment opportunities for the workforce of the country. The following are some of the initiatives to promote the local manufacturing of RE technologies in Zimbabwe:

Standard specifications: The regulator will ensure that the RE products and equipment available in the market adhere to the designed standards and specifications by ZERA and SAZ.

Domestic content requirements: A certain percentage of the RE technology systems installed in the country are mandated to have domestic content. Mandatory local content requirements shall be implemented when there is a sizable demand.

Financial and tax incentives: Financial incentives include providing low-interest sources for project financing, custom duties and tax incentives as provided in the respective legislation (Finance Act, Income Tax Act, Value Added Tax Act and Value Added Tax Regulations, Customs and Excise Duty Act and Customs and Excise General Regulations).

Promoting research and development (R&D) in RE technologies: It is essential to initiate R&D activities related to RE technologies that are pivotal for the long-term sustainability of the industry and developing tailor-made solutions for the Zimbabwean market.

(*Source*: List adapted from Final draft: National renewable energy policy, Republic of Zimbabwe Ministry of Energy and Power Development, 2016.)

CHAPTER 8

Environmental Impact Assessment

THIS CHAPTER PRESENTS A preliminary environment impact assessment as a result of the products and waste material produced from the resource recovery of value-added products. The environmental impact assessment covers the impact of gaseous, liquid and solid pollutants on air, water, soil and human and animal life. An environmental impact is defined as any change to an existing condition of the environment (Figure 8.1).

The nature of the impacts may be categorized in terms of

- *Direction*: Positive or negative

- *Duration*: Long or short term

- *Location*: Direct or indirect

- *Magnitude*: Large or small

- *Extent*: Wide or local

- *Significance*: Large or small

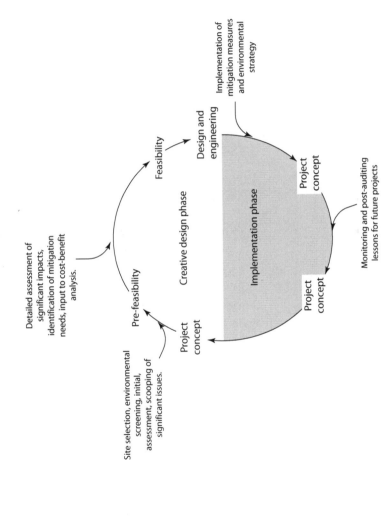

FIGURE 8.1   Environmental impact assessment phases.

8.1 BACKGROUND

Sewage treatment plants are constructed to transform raw sewage into an easier, manageable waste and to retrieve and reuse the treated sewage water. The end products of a treatment plant are sludge and treated sewage water. Both products may contain, in addition to organic biodegradable substances and microorganisms, nonbiodegradable and toxic substances due to the contamination of sewage with industrial wastewaters.

From the environmental standpoint, the most important aspect of a sewage treatment plant is the proposed disposal or use of the sludge and the treated sewage water. The most common adverse environmental effects on coastal waters, connected with disposal or use of the sludge and the treated sewage water, are caused by microbiological contamination, oxygen depletion due to a high load of organic fecal matter, eutrophication caused by nutrients and toxic and nonbiodegradable substances originating mainly from contamination of sewage by industrial waste. Most sewage treatment and disposal processes are a serious source of offensive odor. Improperly constructed or operated sewage treatment plants and improper disposal or use of sludge and treated sewage water may become a most serious public health problem.

8.2 SPECIFIC OBJECTIVES TO BE MET

- Establish a dependable wastewater treatment plant, which is cost-effective and requires minimum maintenance.

- Manage the pathogenic risk inherent in wastewater to meet the effluent discharge standards set by the Standards Association of Zimbabwe.

- Manage the safe disposal of sludge.

8.3 LIKELY SIGNIFICANT ENVIRONMENTAL EFFECTS AND THEIR MITIGATION

The establishment of the wastewater treatment plant will have positive environmental impacts, even if minimum maintenance is applied. It is unlikely that it will pollute the surface waters or soil, or contaminate the aquifers in the area. However, if the wastewater treatment plant is not established, wastewater from the septic tanks of manholes will continue discharging raw sewage into nature, and this will exert negative effects on the local environment and continue to create health problems.

The overall environmental effects of the project will be insignificant. Nevertheless, there is a chance of minor soil erosion incidents, caused by runoff during the wet season of the construction phase. In addition, a minor visual inconvenience can be experienced due to the physical existence of the plant, about 800 m from the town. However, the presence of trees around the wastewater treatment plant will hide the structures.

8.3.1 Site-Specific Factors

The choice of the site is based on the fact that it is the convenient low spot from the town, where wastewater from all houses can reach gravitationally. Actually, it is the only available single valley where the wastewater can be collected gravitationally.

The treated effluent of wastewater treatment will be discharged or used for afforestation purposes, and usually stays dry during the May to November period. In here, the effluent will be further purified by natural factors, such as sun, soil, bacteria and other microorganisms. After that stage, there is no probability that any significant environmental hazards might occur.

8.3.2 Process Technology-Related Effects

The treatment of sewage using Hycura is an environmentally friendly and low-cost technology that can be applicable in rural areas where the availability of skilled labor is scarce.

Compared with other similar wastewater treatment technologies, it is dependable, because of the absence of mechanical equipment, which often can go out of order. The system operates gravitationally. Water pumping requirements are minimal in the plant. On the other hand, the sludge is kept in the anaerobic digesters where methane gas will be generated and captured. There are no odorous gas emissions from the plant. In order to eliminate the release of greenhouse gases in the atmosphere, the generated biogas inside of the digester is trapped as it is produced.

However, there is a chance of mosquito breeding in the open aeration area of the plant. This might create a nuisance to the people, but since mosquitoes do not travel for more than 500 m from their breeding sites, the chance of mosquitoes reaching the town is very low.

8.3.3 Effects Created during Site Preparation and Construction

During the construction phase of the project, which will last for 3–4 months, a few trees of the site will be cut, moderate quantities of earth will be excavated (maximum of 2 m depth) and soil disturbance will take place. If this soil is not utilized for landscaping, during the wet season soil erosion will result at the site. On the other hand, if the excavated soil is haphazardly dumped, this will cover trees and block canals in the downstream direction and create unsightly scenes at the project site.

8.3.4 Effects during Operation

No significant environmental effects will take place during the operation of the plant. The influent wastewater will be treated to a secondary level, as required by the disposal standards. The sand and grease of the influent wastewater will be properly managed. The sludge of the plant will be cleaned and dried for use as a fertilizer.

The accumulated biogas in the anaerobic digester will be trapped in gas tanks. This will eliminate methane gas emissions to the atmosphere. Nevertheless, there is a low probability that the relief valve of the digester might be clogged and the pressure in the anaerobic chamber increased, and this may cause a lot of greenhouse gases to be released into the atmosphere.

8.4 MONITORING AND SUPERVISION PROGRAMS

8.4.1 Monitoring

The sewage treatment plant does not need full-time attendance, as it is self-operational, but it will be subjected to continuous monitoring. Actually, the only need is to test for the final effluents and report any changes that might be caused by outside factors. Monitoring activities will also cover the odor, mosquito and other nuisances that might result at the plant site. This will ensure that all of the mitigation measures are within the safe limit and do not have environmental effects.

8.4.2 Supervision

The relevant municipalities will be in charge of the supervision of the sewage treatment plant operation. The operational manual of the plant gives detailed instructions as to what to do weekly, monthly and yearly. The only attention the plant needs is the cleanup of the grease and sand trap. The operator has to check the grease and sand trap weekly and remove the floating grease and submerged sand. Testing must be done at least once a month to check the pH and biological oxygen demand (BOD).

The process does not need full-time attendance. For the removal of grease and sand, one laborer can periodically open the manhole cover and remove the floating grease and other floating material, as well as remove the sand basket and empty it. Also, proper periodic checkups will be done for the gas relief valves to ensure their proper functioning. Once every 2 years, the digesters will be cleaned from the accumulated sludge, which will be dried on site and then used for agriculture.

8.5 SITE PREPARATION AND VEGETATION CLEARANCE

8.5.1 Impacts

Site clearance and construction practices generally mean the removal of existing vegetation.

These practices remove protective plant cover and expose the soil to erosive surface runoff during heavy rainfall. The inappropriate disposal of the cleared vegetation could lead to burning on site and associated negative impacts on local air quality.

8.5.2 Mitigation

1. Vegetation site clearance should be phased and the project site cleared as the need arises, as opposed to the practice of clearing the entire site in a single major clearance exercise. This will help to minimize the amount of bare or exposed soil present at the site, and thereby help reduce the risk of soil erosion during heavy rains and flash flooding.

2. Areas of exposed soil should be replanted with grass as soon as possible after construction to help mitigate against flash flood soil erosion.

3. To reduce the amount of organic waste generated by the project, small- to medium-sized branches and bits of vegetation may be placed through an on-site commercial wood chipper. The resulting wood chips may then be recycled (i.e., on or off site) as soil cover and similar soil amendment undertakings associated with either project-related or non-project-related landscaping. With regard to a practical suggestion for recycling larger and harder tree trunks and branches, the latter may be made available to local charcoal burners.

8.6 IMPACT: NOISE POLLUTION

Site clearance and construction of the proposed development necessitates the use of heavy equipment to carry out the nature of the job. This equipment includes bulldozers and backhoes.

They possess the potential to have a direct negative impact on the environment.

8.6.1 Mitigation

- Use equipment that has low noise emissions as stated by the manufacturers.

- Use equipment that is properly fitted with noise reduction devices such as mufflers.

- Operate noise-generating equipment during regular working hours (e.g., 7:00 a.m. to 7:00 p.m.) so as to reduce the potential of creating a noise nuisance during the night.

- Construction workers operating equipment that generates noise should be equipped with noise protection. A guideline is that a worker operating equipment generating noise of ≥ 80 dBA (decibels) continuously for 8 hours or more should use earmuffs. Workers experiencing prolonged noise levels of 70–80 dBA should wear earplugs.

8.6.2 Impact: Water Quality

Removal of the vegetation can result in high suspended sediment concentrations in the runoff from the site, during construction. Fortunately, the majority of the earthworks are depressions, and hence the storm water will be naturally retained in the basins.

8.6.3 Mitigation

Surface runoff will be controlled by temporarily beaming the outlet of the significant storm water features to provide some detention behind the berms.

8.7 IMPACT: AIR QUALITY

Site preparation and construction have the potential to have a two-fold direct negative impact on air quality. The first impact is air pollution generated from the construction equipment and transportation. The second is from fugitive dust from site and access roads, cleared areas and raw materials stored on site. Fugitive dust has the potential to affect the health of construction workers, the resident population and the vegetation. The mitigation that can be considered includes

- Site roads should be dampened every 4–6 hours or within reason to prevent a dust nuisance, and on hotter days, this frequency should be increased.

- The access roads (unpaved sections) through the dusty land should also be wetted and the sections of the road monitored so that any material falling on it as a result of the construction activities can be removed.

- Minimize cleared areas to those that need to be used.

- Cover or wet construction materials such as marl to prevent a dust nuisance.

- Where unavoidable, construction workers working in dusty areas should be provided and fitted with N95 respirators.

8.8 IMPACT: EMPLOYMENT

During this phase, an average of approximately 47 persons will be employed. This has the potential to have a significant positive impact. There is no mitigation required here.

8.9 IMPACT: SOLID WASTE GENERATION

During this construction phase of the proposed project, solid waste generation may occur mainly from two points:

- From the construction campsite

- From construction activities such as site clearance and excavation

8.9.1 Mitigation

1. Skips and bins should be strategically placed within the campsite and construction site.

2. The skips and bins at the construction campsite should be adequately designed and covered to prevent access by vermin and minimize odor.

3. The skips and bins at the construction site should be adequately covered to prevent a dust nuisance.

4. The skips and bins at both the construction campsite and construction site should be emptied regularly to prevent overfilling.

5. Trees that are removed from the proposed site may be given to local persons to be used for lumber or for charcoal burning.

8.10 IMPACT: ODOR

A wastewater treatment plant carries a risk of becoming an odor nuisance if proper buffers between the treatment units and existing populations are not provided. A buffer of at least 100 m has been provided on all boundaries per the Standards Association of Zimbabwe recommendations. Additionally, the perimeter of the proposed site will be vegetated with trees and plants of varying heights, thereby forming a windbreaker.

8.10.1 Mitigation

1. Monitor and ensure that influent sulfate levels are below 240 mg/L.

2. Ensure that the pond series have adequate water flow to reduce the potential of odor formation.

8.11 STATUTORY INSTRUMENTS

1. Ministry of Mines and Minerals

2. Environmental Management Agency

3. Environment Management Act (Chapter 20:27)

It provides for the sustainable management of natural resources, the protection of the environment and the prevention of pollution and environmental degradation. The act also repeals the following former acts:

- Natural Resources Act (Chapter 20:13)

- Atmospheric Pollution Prevention Act (Chapter 20:33)

- Hazardous Substances and Article Act (Chapter 15:05)

4. Mines and Minerals Act

The Mine and Minerals Act defines mining as intentionally digging or excavating minerals. It furthermore defines a mineral as any substance, in solid or liquid form, occurring naturally in or on the earth, formed by ore subject to geological processes. Any process requiring this should have an Environmental Impact Assessment (EIA).

5. Gold Act (Chapter 21:03)

6. Forest Act

7. Wildlife Act

8. Explosives Substances Act

From the mentioned acts, this project will be governed by the Natural Resources Act, Environmental Management Act, Explosives Act and Atmospheric Pollution Prevention Act.

CHAPTER 9

Conclusion and Recommendations

9.1 CONCLUSIONS

Resource recovery from municipal plants provides economic, social and financial benefits for municipal sewage plants, as well as their surrounding communities. The treated effluent can be recovered and used as irrigation water boosting the agriculture sector. At the same time, the biogas produced provides an alternative and clean source of renewable energy that also helps to combat climate change in developing countries. Furthermore, biosolids (anaerobic digestate) recovered from the process provide an odor-free and NPK-rich biofertilizer that can be used by farmers to improve the soil fertility. Three products, water, biogas and biosolids, can be efficiently harnessed by using the bioaugmentation technology and adoption of biocatalysis, such as Hycura, in municipal plants for developing countries.

Case studies for the Chitungwiza, Firle and Crow borough municipal sewage plants indicated that there is potential to harness biogas for energy usage at the plants. However, the following issues must be addressed:

- Optimizing the operation of the municipal sewage treatment plants

- Harnessing the biogas from the municipal sewage plants so that it is economically viable for meeting the sewage plants' needs, and excess can be sold to the national grid to meet the energy deficit

9.2 RECOMMENDATIONS

As for sewage biogas energy harnessing, the following are recommended:

1. Cooperation or collaboration links should be established between the energy regulatory authorities, research institutes, relevant ministries and municipal sewage plants for harnessing biogas for reuse and transfer to the national grid.

2. In the new municipal sewage plants, new proven technologies should be investigated and possibly adopted, such as the Hycura technology, for optimal biogas production.

Glossary of Terms

Anaerobic digestion: This is the process by which organic wastes are biologically transformed in the absence of oxygen.

Batch system: Microbial culture produced by inoculating a closed-culture vessel containing a single batch of medium. Conditions within batch culture change dynamically with time.

Biocatalyst: A biological microorganism used to speed up a reaction.

Biogas: Gas produced from digestion of any organic material.

Bionutrients removal: A process for removal of nitrogen and phosphorous in wastewater.

Biosolids: Solids obtained from the anaerobic digestion of organic material.

Hycura: A solid biocatalyst that is a combination of several enzymes and bacteria used in wastewater treatment.

Mesophilic conditions: These are conditions whereby an organism can grow in the temperature range 20°C–45°C.

Methanogenesis: The process by which sewage is degraded to form biomethane. This is the last stage of the anaerobic digestion involving biomethane-producing microorganisms called methanogens.

Sewage: Wastewater from domestic and municipal uses.

Sewage effluent: Treated sewage at a sewage treatment plant.

Sewage influent: Untreated sewage at a sewage treatment plant.

Sludge: The semiliquid material that is removed from a wastewater treatment system as an end product of the treatment process.

Thermophilic conditions: Describes conditions whereby an organism can grow at high temperatures $\geq 45°C$.

Wastewater: Any water whose set quality parameters have been altered.

Wastewater treatment plant: Any structure, thing or process used for physical, chemical, biological or radiological treatment of wastewater.

References

Abdalla, K. Z. and Hammam, G. (2014) Correlation between biochemical oxygen demand and chemical oxygen demand for various wastewater treatment plants in Egypt to obtain biodegradability indices, *International Journal of Applied Sciences: Basic and Applied Research*, 13 (1), pp. 42–48.

African Water Facility, Chitungwiza. (2009) Water and sanitation rehabilitation project appraisal report, Mobilising Resources for Water in Africa, pp. 1–18.

Ahn, K., Song, K., Cho, E., Cho, J., Yun, H., Lee, S., and Kim, J. (2003) Enhanced biological phosphorous and nitrogen removal using a sequencing anoxic/anaerobic membrane bioreactor (SAM) process, *Desalination*, 157, pp. 345–352.

Ahsan, S., Rahman, S., Kaneco, S., Katsumata, S., Suzuki, T., and Onta, K. (2005) Effect of temperature on wastewater treatment with natural and waste materials, *Clean Technology Environment Policy*, 7, pp. 198–202.

Al Awadhi, M. A. A. N. (2013) Beneficial re-use of treated sewage effluent, Director of Sewage Treatment Plant Department, Dubai Municipality.

APHA (America Public Health Association). (2005) *Standard Methods for the Examination of Water and Wastewater*, 21st Edition, America Public Health Association, American Water Works Association, Water Environment Federation, Washington, DC.

Apte, S. S., Apte, S. S., Kore, V. S., and Kore, V. S. (2011) Chloride removal from wastewater by bio-sorption with the plant biomass, *Universal Journal of Environment Research and Technology*, 1 (4), pp. 416–422.

Arthur, R. and Brew-Hammond, A. (2010) Potential biogas production from sewage bio-solids: A case study of the sewage treatment plant at Kwame Nkrumah University of Science and Technology, Ghana, *International Journal of Energy and Environment*, 1 (6), pp. 1009–1016.

Attiogbe, F. K., Glover-Amengor, M. G., and Nyadziehe, K. T. (2007) Correlating biochemical and chemical oxygen demand of effluents—A case study of selected industries in Kumasi, Ghana, *West African Journal of Applied Ecology*, 1, pp. 1–11.

Babaee, A. and Shayegan, J. (2011) Effect of organic loading rate (OLR) on production of methane from anaerobic digestion of vegetable waste, World Renewable Energy Congress, Bioenergy Technology, 8–13 May, Linkoping, Sweden.

Barker, P. S. and Dold, P. L. (1996) Denitrification behaviour in biological excess phosphorous removal activated bio-solids systems, *Water Research*, 30 (4), pp. 769–780.

Bharathiraja, B., Chakravarthy, M., Ranjith, R., Kumar, D., Yuvaraj, J., Jayamuthunagai, R., Kumar, P., et al. (2014) Biodiesel production using chemical and biological methods—A review of process, catalyst, acyl acceptor, source and process variables, *Renewable and Sustainable Energy Reviews*, 38, pp. 368–382.

Cail, R. G., Barford, J. P., and Linchacz, R. (1986) Anaerobic digestion of wool scouring wastewater in digester operated semi-continuously for biomass retention, *Agricultural Wastes*, 18, pp. 27–38.

CH2MHILL. (2011) Niagara region biosolids management master plan, Final report, 2011.

Coelho, S. T., Velazquez, S. M. S. G., Martins, O. S., and Abreu, F. C. (2006a) Biogas from sewage treatment used to electric energy generation, by a 30 kW (ISO) microturbine, World Bioenergy Conference and Exhibition, 30 June, Jonkoping, Suecia.

Coelho, S. T., Velazquez, S. M. S. G., Silva, O. C., Pecora, V., and Abreu, F. C. (2006b) The production of sewage biogas and its use for energy generation, World Bioenergy Conference and Exhibition, 30 June, Jonkoping, Suecia.

Coppen, J. (2004) Advanced wastewater treatment system, Courses ENG 411 and 412 Research Project.

Daelman, M. R. J., van Voorthuizen, E. M., van Dongen, U. G. J. M., Volcke, E. I. P., and van Loosdrecht, M. C. M. (2012) Bio-methane emission during municipal wastewater treatment, *Water Research*, 46, pp. 3657–3670.

Davis, R., Aden, A., and Pienkos, P. (2011) Techno-economic analysis of autotrophic microalgae for fuel production, *Applied Energy*, 88, pp. 3524–3531.

Desai, S. T., Palled, V., and Mathad, R. (2013) Performance evaluation of fixed doom type biogas plant for solid state digestion of cattle dung, *Karnataka Journal of Agricultural Sciences*, 26 (1), pp. 103–106.

Doyle, M. P. and Padhye, V. S. (1989) *Escherichia coli*. In *Food-Borne Bacterial Pathogens*, Doyle, M. P., Ed., Marcel Dekker, New York.

Driessen, W., Yspeert, P., Yspeert, Y., and Vereijken, T. (2000) Compact combined anaerobic and aerobic process for the treatment of industrial effluent, *Environmental Forum*, pp. 1–10.

Duncan, A. C. (1970) Two stage anaerobic digestion of hog wastes, Master of science thesis, University of British Columbia, 1970.

Dzvene, D. K. (2013) An investigation on the effectiveness of actizyme bacteria to treat organic waste at Tripple C Pigs (Colcom), Bachelor of Science Honours Degree in Environmental Sciences and Technology, School of Agricultural Sciences and Technology, Chinhoyi University of Science and Technology, November 2013.

Ecolab. (2006) Material safety data sheet, Hycura Compost Accelerator, 903100, pp. 1–3.

El-Fadel, M. and Massoud, M. (2001) Methane emissions from wastewater management. *Environmental Pollution*, 114, pp. 177–185.

Evanylo, G. K. (2009) Agricultural land application of bio-solids in Virginia: Production and characteristics of bio-solids, *Virginia Cooperative Extension*, Publication 425–301, pp. 1–6.

Gebrezgabher, S. A., Meuwissen, M. P. M., and Oude Lansink, A. G. J. M. (2010a) Cost of producing biogas at dairy farms in the Netherlands, *International Journal of Food System Dynamics*, 1, pp. 26–35.

Gebrezgabher, S. A., Meuwissen, M. P. M., Prins, B. A. M., and Oude Lausink, G. J. M. (2010b) Economic analysis of anaerobic digestion—A case of green power biogas plant in the Netherlands, *NJAS—Wageningen Journal of Life Sciences*, 57 (2), pp. 109–115.

Gomes, A. C., Silva, L., Simoes, R., Canto, N., and Alburquerque, A. (2013) Toxicity reduction and biodegradability enhancement of cork processing wastewaters by ozonation, *Water Science and Technology*, 68 (10), pp. 2214–2219.

Gorveno, J., Kiepper, B., Magbanua, B., and Adams. T. (2000) Characterisation and quantification of Georgia's municipal bio-solids production and disposal, University of Georgia, College of Agriculture and Environmental Sciences, Department of Biological and Agricultural Engineering.

Gupta, V. K., Ali, I., Saleh, T. A., Nayak, A., and Agarwal, S. (2012) Chemical treatment technologies for waste-water recycling—An overview, *Royal Society of Chemistry Advances*, 2, pp. 6380–6388.

Helmer, R. and Hespanhol, I., Eds. (1999) *Water Pollution Control. A Guide to the Use of Water Quality Management Principles*. UNEP: United Nations Environment Programme.

Hesnawi, R. M. and Mohamed, R. A. (2013) Effect of organic waste source on methane production during thermophilic digestion process, *International Journal of Environmental Science and Development*, 4 (4), pp. 435–437.

Hospido, A., Moreira, M. T., Martin, M., Rigola, M., and Feijoo, M. (2005) Environmental evaluation of different treatment processes for bio-solids from urban wastewater treatments: Anaerobic digestion versus thermal processes, Urban Wastewater Treatments, Waste in LCA. *The International Journal of Life Cycle Assessment*, 10 (5), pp. 336–345.

Hu, B. L., Shen, L. D., Xu, X. Y., and Zheng, P. (2011) Anaerobic ammonium oxidation in different natural ecosystems, *Biochemical Society Transactions*, 39 (6), pp. 1811–1816.

Huang, X., Xiao, X., and Shen, Y. (2010) Recent advances in membrane bioreactor technology for wastewater treatment in China, *Frontiers of Environmental Science and Engineering*, 4 (3), pp. 245–271.

Johannesson, G. H. (1999) Sewage characteristics and evaluation of P availability under greenhouse conditions, Master of science thesis, University of Guelph.

Kakore, N. (2014). Methane power plant for city. http://www.herald.co.zw/methane-power-plant-for-city/.

Kampas, P., Parsons, S. A., Pearce, P., Ldoux, S., Vale, P., Churchley, J., and Cartmell, E. (2007) Mechanical bio-solids disintegration for the production of carbon source for biological nutrient removal, *Water Research*, 41 (8), pp. 1734–1742.

Kardas, L., Jahasz, A., Palko, G. Y., Olah, J., Bakacs, K., and Zaray, G. Y. (2011) Comparing of mesophilic and thermophilic anaerobic fermented sewage sludge based on chemical and biochemical tests. *Applied Ecology and Environmental Research*, 9 (3) 293–302.

Kargbo, D. M. (2010) Biodiesel production from municipal sewage bio-solids, *Energy Fuels*, 24, pp. 2791–2794.

Kaosol, T. and Sohgrathok, N. (2012) Enhancement of biogas production potential for anaerobic co-digestion of wastewater using decanter cake, *American Journal of Agriculture and Biological Sciences*, 7 (4), pp. 494–502.

Kozani, S. J. (2014) Basics of the biogas production process. Workshop at the Institute of Biotechnology, Berlin Institute of Technology, Germany.

Kraume, M., Bracklow, U., Vocks, M., and Drews, A. (2005) Nutrients removal in MBRs for municipal wastewater treatment, *Water Science Technology*, 51, pp. 391–402.

Lai, T. M., Shin, J., and Hur, J. (2011) Estimating the biodegradability of treated sewage samples using synchronous fluorescence spectra, *Sensors*, 11, pp. 7382–7394.

Lang, N. L. and Smith, S. R. (2007) Influence of soil type, moisture content and bio-solids application on the fate of *Escherichia coli* in agricultural soil under controlled laboratory conditions. *Journal of Applied Microbiology*, 103, p. 2122–2131.

Lang, N. L., Bellet-Travers, M. D., and Smith, S. R. (2007). Field investigation on the survival of *Escherichia coli* and presence of other enteric micro-organisms in bio-solids amended agricultural soil. *Journal of Applied Microbiology*, 103, p. 1868–1882.

Lee, A. H. and Hamid, N. (2014) BOD:COD ratio as an indicator for pollutants leaching from landfill, *Journal of Clean Energy Technologies*, 2 (3), pp. 263–266.

Lettinga, G. (1995) Anaerobic digestion and wastewater treatment systems, *Antonie van Leeuwenhoek*, 67, pp. 3–28.

Mace, S. and Mata-Alvarez, J. (2002) Utilisation of SBR technology for wastewater treatment: An overview, *Industrial and Engineering Chemistry Research*, 41, pp. 5539–5553.

Mahmoud, N. (2002) Anaerobic pre-treatment of sewage under low temperature (15 °C) conditions in an integrated UASB-digester system, PhD thesis, Wageningen University, Wageningen, The Netherlands.

Mahmoud, N. (2011) Wastewater characteristics, Ecological Sanitation Training Course, Switch Project, 25–27 January.

Mahmoud, N., Zeeman, G., Gijzen, H., and Lettinga, G. (2003) Anaerobic sewage treatment in a one stage UASB and a combined UASB-digester system, Seventh International Water Technology Conference, Cairo, Egypt, 28–30 March.

Malik, D. S. and Bharti, U. (2009) Biogas production from sludge of sewage treatment plant at Haridwar (Uttarakhand), *Asian Journal of Experimental Sciences*, 23 (1), pp. 95–98.

Manyuchi, M. M., Ikhu-Omoregbe, D. I. O., and Oyekola, O. O. (2015) Acti-zyme biochemical properties: Potential for use in anaerobic sewage treatment co-generating biogas, *Asian Journal of Science and Technology*, 6 (3), 1152–1154.

Manyuchi, M. M., Kadzungura, L., and Boka, S. (2013) Pilot studies for vermifiltration of 1000 m³/day of sewage wastewater, *Asian Journal of Engineering and Technology*, 1 (1), pp. 13–19.

Manyuchi, M. M. and Phiri, A. (2013) Application of the vermifiltration technology in sewage wastewater treatment, *Asian Journal of Engineering and Technology*, 1 (4), pp. 108–113.

Masvingise, M. and Bare, K. (2010). Harare mulls power station takeover. http://www.financialgazette.co.zw/harare-mulls-power-station-takeover/.

McBride, M. B. (2003) Toxic metals in sewage sludge amended soils: Has promotion of beneficial use discounted the risks? *Advances in Environmental Research*, 8 (1), pp. 5–19.

Mes, T. Z. D. de; Stams, A. J. M., and Zeeman, G. (2003) Methane production by anaerobic digestion of wastewater and solid wastes. In *Biomethane and Biohydrogen. Status and Perspectives of Biological Methane and Hydrogen Production*, Reith, J. H., Wijffels, R. H., and Barten, H., Eds., Dutch Biological Hydrogen Foundation, The Hague, The Netherlands, pp. 58–102.

Mittal, A. (2011) Biological wastewater treatment. *Water Today*, August, pp. 32–44.

Mohan, S. V., Rao, N. C., Prasad, K. K., Madhavi, B. T. V., and Sharma, P. N. (2005) Treatment of complex chemical wastewater in a sequencing batch reactor (SBR) with an aerobic suspended growth and configuration, *Process Biochemistry*, 40, pp. 1501–1508.

Molwantwa, J. B. (2002) The hydrolysis of primary sewage sludge under bio-sulphidogenic conditions, Master of science thesis, Rhodes University.

Mukumba, P., Makaka, G., Mamphwel, S., and Misi, S. (2013) A possible design and justification for a biogas plant at Nyazura Adventist High School, Rusape Zimbabwe, *Journal of Energy in Southern Africa*, 24 (4), 12–21.

Muserere, S. T., Hoko, Z., and Nhapi, I. (2014) Characterisation of raw sewage and performance assessment of primary settling tanks at Firle sewage treatment works, Harare, Zimbabwe, *Physics and Chemistry of the Earth*, 67–69, pp. 226–235.

Nazaroff, W. W. and Alvarez-Cohen L. (2013) Anaerobic digestion of wastewater bio-solids, Section 6.E.3.

Neczaj, E., Grosser, A., and Worwag, M. (2013) Boosting production of bio-methane from sewage sludge by addition of grease trap sludge, *Environmental Protection Engineering*, 39 (2), pp. 125–133.

Nhapi. I. (2009) The water situation in Harare, Zimbabwe: A policy and management problem, *Water Policy*, 11, pp. 221–235.

Nhapi, I. and Gijzen, H. (2002) Wastewater management in Zimbabwe, *Sustainable Environmental Sanitation and Water Services, 28th Conference*, Calcutta, India, pp. 181–184.

Ozmen, P. and Aslaunzadeh, S. (2009) Biogas production from municipal waste mixed with different portions of orange peel, MSc thesis, Applied Biotechnology, University of Boras, School of Engineering.

Pabsch, H. and Wendland, C. (2013) Sewage bio-solids treatment, Lesson B5, Efficient management of wastewater.

Palanisamy, K. and Shamsuddin, A. H. (2013) Biogas to energy potential from sewage treatment plant in Malaysia, 1st Malaysia National Sewage Conference, Sewerage Malaysia-Setting New Frontier Technology, June.

Popa, P., Timofti, M., Voiculesai, M., Dragan, S., Trif, C., and Georgescu, L. P. (2012) Study of physicochemical characteristics of wastewater in an urban agglomeration in Romania, *The Scientific World Journal*, article ID 549028. http://dx.doi.org/10.1100/2012/549028.

Powell, B. and Lundy, J. (2007) Hycura agricultural and municipal wastewater treatment, Environment Depot Canada, pp. 1–23.

Reali, M. A., Campos, J. R., and Penetra, R. G. (2001) Sewage treatment by anaerobic biological process associated with dissolved air floatation, *Water Science and Technology*, 43 (8), pp. 91–98.

Republic of Zimbabwe Ministry of Energy and Power Development. Final draft: National Renewable Energy Policy, 2016.

Rich, J. J., Dale, O. R., Song, B., and Ward, B. B. (2008) Anaerobic ammonium oxidation (Anammox) in Chesapeake Bay sediments, *Microbiology Ecology*, 55, pp. 311–320.

Rodriguez-Chiang, L. M. and Dahl, O. P. (2015) Effect of innoculum to substrate ratio on the methane potential of microcrystalline cellulose production wastewater, *Bioresources*, 10 (1), pp. 898–911.

Schreckenberger, P. C. and Blazevic, D. J. (1974) Rapid methods for biochemical testing of anaerobic bacteria, *Applied Microbiology*, 28 (5), pp. 759–762.

Segal, K. and Potter, B. (2008) Identification of unknown bacteria using a series of biochemical tests, *Microbiology* 202, pp. 1–10.

Shin, J., Sang, L., Jung, J., Chung, Y., and Noh, S. (2005) Enhanced COD and nitrogen removals for the treatment of swine wastewater by combining submerged membrane bioreactor (MBR) and anaerobic upflow bed filter (AUBF) reactor, *Process Biochemistry*, 40, pp. 3769–3776.

Singh, S. and Singh, K. N. (2010) Physicochemical analysis of sewage discharged into Varuna River at Varamasi, *Current World Environment*, 5 (1), pp. 201–203.

Tas, D. O., Karahan, O., Insel, G., Ovez, S., Orhon, D., and Spanjers, H. (2009) Biodegradability and denitrification potential of settleable chemical oxygen demand in domestic wastewater, *Water Environment Research*, 81 (7), pp. 715–727.

Tshuma, A. C. (2010) Impact of Acti-zyme compound on water quality along mid-mupfuure Catchment-Chegutu, BSc honors thesis, Geography and Environmental Studies, Midlands State University, Zimbabwe.

USEPA (U.S. Environmental Protection Agency). (2012) Financing anaerobic digestion projects.

USEPA (U.S. Environmental Protection Agency). (2013) Emerging technologies for wastewater treatment and in-plant wet weather management, 2 July.

Vashist, H., Sharma, D., and Gupta, A. (2013) A review of commonly used biochemical tests for bacteria, *Innovare Journal of Life Sciences*, 1 (1), pp. 1–7.

Vindis, P., Marsec, B., Janzekovic, M., and Cus, F. (2009) The impact of mesophilic and thermophilic anaerobic digestion on biogas production, *Journal of Achievements in Materials and Manufacturing Engineering*, 36 (2), pp. 192–198.

Wei, Y., van Houten, R. T., Borger, A. R., Eikelboom, D. H., and Fan, Y. (2003) Minimisation of excess bio-solids production for biological waste water treatment, *Water Research*, 37, pp. 4453–4467.

Willis, J. and Schafer, P. (2006) Advances in thermophilic anaerobic digestion, Water Environment Foundation, WEFTEC 06, pp. 5378–5392.

Zaher, U., Cheong, P., Wu, B., and Chen, S. (2007) Producing energy and fertilizer from organic municipal solid waste. Project Deliverable No. 1, 26 June 26.

Zhou, H. and Smith. D. W. (2002) Advanced technologies in water and wastewater treatment, *Journal of Environmental Engineering and Science*, 1, pp. 247–264.Insert:

Index

Acetogenesis, 23
Acidogenesis, 23
Aeration zone, 67
Aerobes, 12
Aerobic conditions, 12–13
Agitation, 68–69
Air quality, 100–101
America Public Health Association
   (APHA), 47
Anaerobes, 13
Anaerobic conditions, 13
Anaerobic degradation, 44
Anaerobic treatment, of wastewater,
   15–20
   Hycura
   biochemical properties, 16–17
   kinetics, 19–20
   use, 17–18
   overview, 15–16
Anammox conditions, 13–14
AND moisture analyzer, 59
Anoxic conditions, 13
Anoxic zone, 67
APHA, *see* America Public Health
   Association (APHA)
Atmospheric Pollution Prevention
   Act, 103

Bioaugumentation, 10–12
   anaerobic treatment, 15–20
   Hycura, 16–20
   overview, 15–16

nutrient removal and, 10–12
   primary, 11
   secondary, 11–12
   tertiary, 12
Biodigester
   design, 44–46
   section description, 67–70
   agitation, 68–69
   biogas purification, 70
   carbon-to-nitrogen ratio,
      69–70
   organic loading rate, 70
   pH, 68
   retention time, 68
   temperature, 68
   total solids content, 68
   toxicity, 70
Biofertilizers, 29, 63
Biogas and biosolids recovery, 58–64
   experimental approach, 59
   overview, 58–59
   production, 59–64
   biofertilizers, 63
   carbon dioxide generation,
      61, 63
   high-nitrogen-content
      biosolids, 64
   methane generation, 61
   trace gas generation, 63
   raw sewage sludge, 59
Biogas from sewage sludge, 3, 23–28,
   105–106

benefits, 26–28
 alternative and renewable
  energy source, 26
 employment creation and
  income generation, 28
 reduced dependence on fossil
  fuels, 28
 reduced greenhouse
  emissions, 26, 28
 waste reduction and
  management, 28
 conditions, 23–26
 Hycura bioaugmentation, 26
Biogas production, 43–44, 73–75
Biogas purification, 70
Biological filters, 66
Biological nutrient removal (BNR)
  plant, 65, 67
 aeration zone, 67
 anoxic zone, 67
 clarifier, 67
 fermentation basin, 67
 primary settling tanks, 67
Biological oxygen demand (BOD), 8,
  17, 50–51
Biomethane, 24, 25–26
Bionutrient recovery, 22, 55–58
 overview, 55
 removal coefficients, 55, 57–58
  BOD/TKN ratios, 57
  COD/BOD ratios, 55, 57
  COD/TKN ratios, 57
  COD/TP ratios, 57–58
 treatment with Hycura, 55
Biosolids, 3, 28–32, 105
 benefits, 31–32
  closed nutrient and carbon
   cycle, 31
  income, 31
  reduced odors and flies, 32
 biogas and, 58–64
  experimental approach, 59
  overview, 58–59

production, 59–64
 raw sewage sludge, 59
 characteristics, 29–31
BNR, see Biological nutrient removal
  (BNR) plant
BOD, see Biological oxygen
  demand (BOD)
BOD/TKN ratios, 57
Breakeven point, 35

C/N ratio, see Carbon-to-nitrogen
  (C/N) ratio
Carbon dioxide, 61, 63
Carbon-to-nitrogen (C/N) ratio,
  69–70, 105
Chemical oxygen demand
  (COD), 8, 51
 and BOD ratios, 55, 57
 and TKN ratios, 57
 and TP ratios, 57–58
Chitungwiza plant, resource recovery
  from, 39–42, 105
 biodigester design, 44–46
 biodigester section description,
  67–70
  agitation, 68–69
  biogas purification, 70
  carbon-to-nitrogen ratio,
   69–70
  organic loading rate, 70
  pH, 68
  retention time, 68
  temperature, 68
  total solids content, 68
  toxicity, 70
 biogas and biosolids, 58–64
  experimental approach, 59
  overview, 58–59
  production, 59–64
  raw sewage sludge, 59
 biogas production, 43–44, 73–75
 bionutrient, 55–58
  overview, 55

removal coefficients, 55, 57–58
treatment with Hycura, 55
economic assessment for
recovering biogas, 75–76
energy balance, 75
feasibility determination,
75–76
mass, energy and balance, 76
features, 70–72
overview, 39–42
problem statement, 42
study objectives, 42
process description, 65–67
aeration zone, 67
anoxic zone, 67
biological filters, 66
clarifier, 67
fermentation basin, 67
humus tanks, 66
maturation ponds, 66
primary settling tanks, 66, 67
treated effluent, 46–54
effect of Hycura loadings and
retention time, 47, 49
effect on BOD, 50–51
effect on Cl- ion
concentration, 53
effect on COD, 51
effect on DO, 53
effect on EC, 52
effect on pH, 49
effect on SO42- ion
concentration, 53
effect on TDS, 52
effect on TKN, 50
effect on total *E. coli* and
coliform content, 53
effect on TP, 49–50
effect on TSS, 51–52
experimental approach, 46–47
overview, 46
raw sewage and
characteristics, 47

Chloride/Cl- ion concentration,
10, 53
Clarifier, 67
Closed nutrient and carbon
cycle, 31
COD, *see* Chemical oxygen
demand (COD)
Color and odor, 9
Conventional municipal sewage
treatment plant, 65–66
biological filters, 66
humus tanks, 66
maturation ponds, 66
primary settling tanks, 66
COP 21, *see* United Nations Climate
Change Conference

Dissolved oxygen (DO), 12, 53
Domestic content requirements, 91

EC, *see* Electrical conductivity (EC)
Economic assessment and biogas
recovering, 75–76
energy balance for electricity, 75
feasibility determination, 75–76
mass, energy and balance, 76
Economic benefits, 33–37
breakeven point, 35
internal rate of return, 34
net present value, 33–34
payback period, 34–35
risks, 35, 37
sensitivity analysis, 35
Effluent, sewage, 21–22
Electrical conductivity (EC), 52
Electricity production, 73, 75
EMA, *see* Environmental Management
Agency (EMA)
Employment
impact, 101
and income, 28
Environment Management
Act, 103

Environmental impact assessment, 93–103
air quality, 100–101
effects and mitigation, 96–98
during operation, 97–98
process technology-related, 96–97
during site preparation and construction, 97
site-specific factors, 96
employment, 101
monitoring, 98
noise pollution, 99–100
mitigation, 100
objectives, 95
odor, 102
mitigation, 102
overview, 95
site preparation and vegetation clearance, 99
impacts, 99
mitigation, 99
solid waste generation, 101–102
mitigation, 102
statutory instruments, 103
supervision, 98
water quality, 100
mitigation, 100
Environmental Management Agency (EMA), 3, 6, 40, 103
Escherichia coli and coliform content, 7, 53
Explosives Substances Act, 103

Fecal coliform content, 7
Fermentation basin, 67
Financial and tax incentives, 91
Firle and Crowborough plants, *see* Chitungwiza plant
Forest Act, 103
Fossil fuels, 28
Fugitive dust, 101

Gender equality and equity, 90
Gold Act, 103
Greenhouse emissions, 26, 28

High-nitrogen-content biosolids, 64
Humus tanks, 66
Hycura, 2, 3, 40, 46, 49, 51
biochemical properties, 16–17
biogas generation, 26
bionutrient removal, 55
kinetics, 19–20
loadings and retention time, 47, 49
use, 17–18
Hydrolysis, 23

Income, employment and, 28
Internal rate of return (IRR), 34

Kyoto Protocol, 41

Maturation ponds, 66
Methane, 61
Methane-rich biogas, 60
Methanogenesis, 24
Mine and Minerals Act, 103
Ministry of Mines and Minerals, 103
Moisture content, 59
Municipal sewage wastewater treatment, 5–14
characteristics, 6–10
chlorides, 10
color and odor, 9
E. coli content, 7
fecal coliform content, 7
organic material, 8
pH, 9
solids, 10
sulfate concentration, 10
temperature, 8
total nitrogen content, 9
total phosphates content, 9
toxic metals and compounds, 10
conditions, 12–14

aerobic, 12–13
anaerobic, 13
anammox, 13–14
anoxic, 13
description, 5–6
nutrient removal using
    bioaugumentation, 10–12
primary, 11
secondary, 11–12
tertiary, 12
resource recovery, 21–32
biogas, 23–28
bionutrient, 22
biosolids, 28–32
effluent, 21–22

Natural Resources Act, 103
Net present value (NPV), 33–34
Noise pollution, 99–100
NPV, *see* Net present value (NPV)
Nutrient removal and
    bioaugumentation, 3,
    10–12, 22
primary, 11
secondary, 11–12
tertiary, 12

Odor, 32, 102
Organic loading rate, 70
Organic material in sewage
    wastewater, 8
BOD, 8
COD, 8

Payback period, 34–35
pH, 9, 49, 68
Poverty eradication and employment
    creation, 90–91
Primary settling tanks, 66, 67
Primary sewage treatment, 11

R&D, *see* Research and
    development (R&D)

RE, *see* Renewable energy (RE)
RE policy, 89–90
accessibility, 90
affordability, 90
development, 89
gender equality, 90
national provisions, 90–91
poverty eradication and
    employment creation,
    90–91
promote local manufacturing, 91
sustainability, 89–90
Renewable energy (RE), 43
Research and development
    (R&D), 91
Resource recovery
from Chitungwiza, Firle and
    Crowborough plants,
    39–87
biodigester design, 44–46
biogas and biosolids recovery,
    58–64
biogas production, 43–44,
    73–75
bionutrient recovery, 55–58
economic assessment, 75–76
features, 70–72
overview, 39–42
process description, 65–70
recovery of treated effluent,
    46–54
from municipal plants, 1–4
biogas, 23–28
bionutrient, 22
biosolids, 28–32
description, 1–2
motivation, 2–4
treated effluent, 21–22
RE policy for, 89–90
accessibility, 90
affordability, 90
development, 89
gender equality, 90

national provisions, 90–91
sustainability, 89–90
Retention time, 68

Screening stage, 65
SE4ALL program, 90
Secondary sewage treatment, 11–12
Sensitivity analysis, 35
Sewage sludge, 40, 95
  biogas and, 3, 23–28
  benefits, 26–28
  conditions, 23–26
  Hycura bioaugmentation, 26
  biosolids and, 3, 28–32
  benefits, 31–32
  characteristics, 29–31
  raw, 59
Sewage solids, 10
SI, *see* Statutory instruments (SI)
SI 274 of 2004, 3
SO42- ion concentration, 53
Solid waste generation, 101–102
Standards and specifications, 91
Standards Association of Zimbabwe,
  95, 102
Statutory instruments (SI), 103
Sulfate concentration, 10

TDS, *see* Total dissolved solids (TDS)
Temperature, 8, 68
Tertiary sewage treatment, 12
TKN, *see* Total Kjeldahl
  nitrogen (TKN)
Total dissolved solids (TDS), 52
Total Kjeldahl nitrogen (TKN), 17,
  22, 50
Total nitrogen content, 9
Total phosphates (TP) content, 9, 17,
  22, 49–50
Total solids (TS) content, 68
Total suspended solids (TSS), 17,
  51–52
Toxic metals and compounds, 10

Toxicity, 70
TP, *see* Total phosphates (TP) content
Trace gas, 63
Treated effluent recovery, 46–54
  effect of Hycura loadings and
    retention time, 47, 49
  effect on BOD, 50–51
  effect on Cl- ion concentration, 53
  effect on COD, 51
  effect on DO, 53
  effect on EC, 52
  effect on pH, 49
  effect on SO42- ion
    concentration, 53
  effect on TDS, 52
  effect on TKN, 50
  effect on total *E. coli* and coliform
    content, 53
  effect on TP, 49–50
  effect on TSS, 51–52
  experimental approach, 46–47
  overview, 46
  raw sewage and characteristics, 47
TS, *see* Total solids (TS) content
TSS, *see* Total suspended solids (TSS)

U.S. Environmental Protection
  Agency (USEPA), 75
United Nations Climate Change
  Conference, 41
USEPA, *see* U.S. Environmental
  Protection Agency
  (USEPA)

Vegetation site clearance and
  construction, 99
  impacts, 99
  mitigation, 99
Volatile matter content, 59

Waste reduction and management, 28
Water quality, 100
Wildlife Act, 103

Printed and bound by CPI Group (UK) Ltd, Croydon, CR0 4YY

22/10/2024

01777627-0004